Sheffield Hallam University
Learning and IT Services
Adsetts Centre City Campus
Sheffield   S1 1WB

The Science an

102 095 642 9

D1333880

The Science and Commerce of Whisky

Sheffield Hallam University
Learning and Information Service
WITHDRAWN FROM STOCK

Ian Buxton
Brollachan Ltd, UK
Email: ian@brollachan.com

and

Paul S. Hughes
International Centre for Brewing and Distilling,
Heriot-Watt University, UK
Email: p.s.hughes@hw.ac.uk

RSCPublishing

ISBN: 978-1-84973-150-8

A catalogue record for this book is available from the British Library

© Buxton & Hughes, 2014

All rights reserved

Apart from fair dealing for the purposes of research for non-commercial purposes or for private study, criticism or review, as permitted under the Copyright, Designs and Patents Act 1988 and the Copyright and Related Rights Regulations 2003, this publication may not be reproduced, stored or transmitted, in any form or by any means, without the prior permission in writing of The Royal Society of Chemistry or the copyright owner, or in the case of reproduction in accordance with the terms of licences issued by the Copyright Licensing Agency in the UK, or in accordance with the terms of the licences issued by the appropriate Reproduction Rights Organization outside the UK. Enquiries concerning reproduction outside the terms stated here should be sent to The Royal Society of Chemistry at the address printed on this page.

The RSC is not responsible for individual opinions expressed in this work.

Published by The Royal Society of Chemistry,
Thomas Graham House, Science Park, Milton Road,
Cambridge CB4 0WF, UK

Registered Charity Number 207890

Visit our website at www.rsc.org/books

SHEFFIELD HALLAM UNIVERSITY

641.252
BU

ADSETTS LEARNING CENTRE

Preface

More than a century has now elapsed since J. A. Nettleton published his magisterial *The Manufacture of Whisky & Plain Spirit*, a comprehensive survey that remained the standard work for many subsequent decades. All who work in this field are conscious of his shadow.

Since then, there have been only two significant texts of note on this important subject, both of which are now out of print – and prohibitively expensive where they can be found. However, in any event, the world of whisky has evolved in remarkable and unexpected ways over the ten years since the last major reference work (*Whisky: Technology, Production and Marketing*. Ed. Inge Russell, 2003) appeared.

In that time, whisky has refreshed its appeal to a new generation of young consumers who had previously rejected it in favour of other distilled spirits; Scotch whisky producers in particular have embarked on the largest expansion of production seen in living memory; a burgeoning craft distilling sector has emerged, challenging many industry orthodoxies and innovating to spectacular effect and scientific knowledge about the processes underlying distilling and maturation has greatly increased.

With all that in mind we felt that a fresh look at the subject was timely. Moreover, we sought to place the subject in a historical and cultural context, without which we believed whisky's richness and

The Science and Commerce of Whisky
By Ian Buxton and Paul S. Hughes
© Buxton & Hughes, 2014
Published by the Royal Society of Chemistry, www.rsc.org

depth could not be fully appreciated. *The Science and Commerce of Whisky* is thus an attempt to revisit old ground and to offer new perspectives.

As the poet would have it "Take off your dram!"

Paul S Hughes
Ian Buxton

Contents

The Science and Commerce of Whisky
By Ian Buxton and Paul S. Hughes
© Buxton & Hughes, 2014
Published by the Royal Society of Chemistry, www.rsc.org

Acknowledgement

I would like to thank the following individuals and organisations for their help: Darren Whitmer (Brown-Forman Cooperage, Louisville, KY), Dr Gordon Steele and Dr John Conner (Scotch Whisky Research Institute), Dr Dennis Watson (Pernod-Ricard), Alan Barclay (Diageo), Dr Monica Lee (Wageningen University) and family and friends who have had to put up with me whilst I undertake this project. Finally a special thanks to Paul Wood (Heineken) for introducing me to Scotch whisky in the first place in a bar called Lemmy's, incongruously located in Leiden.

Paul S. Hughes

I would like to thank the following: Charles Maclean for reading a late draft of Chapter 2 and his helpful comments on it; Anthony Wills (Kilchoman Distillery), Anssi Pysing (Teerenpeli Distillery), Ian Chang and Dr James Swan (Kavalan Distillery), Dr Nicholas Morgan (Diageo plc), Glen Barclay (Scotch Whisky Association), Jason Craig (Cutty Sark) and Ken Grier (The Macallan) for their assistance with respective case studies and, once again, my wife Lindsay for her patience and forbearance during the creation of another book. Any errors and omissions remain entirely my responsibility.

Ian Buxton

The Science and Commerce of Whisky
By Ian Buxton and Paul S. Hughes
© Buxton & Hughes, 2014
Published by the Royal Society of Chemistry, www.rsc.org

CHAPTER 1

Whisky's Historical Development

1.1 ALCHEMY AND THE DEVELOPMENT OF MODERN CHEMISTRY

"There is a glut of chemical books, but a scarcity of chemical truths."
John French's preface to his *Art of Distillation* (1651)

It might, perhaps, be more comfortable for the grand narrative of modern science if alchemy – one of the pillars on which distilling is founded – could be quietly ignored or indeed forgotten entirely. The rational mind denies the contribution of mediaeval mystics and the arcane lore of the alchemist, yet both Robert Boyle and Isaac Newton pursued alchemical studies – Newton for more than 25 years with alchemy central to his religious beliefs.

Early works on alchemy contain detailed descriptions of distilling and many illustrations of stills at work. It must be appreciated that these are, at heart, scientific works and that practitioners saw themselves as seekers after the truth, albeit proceeding from an Aristotelian view of a world comprised of four elements: earth, air, fire and water. Distillation represented the route to a 'fifth essence' or a kind of ultra-purified elixir that, in its highest form, could even prolong life. This was the search for the Philosopher's Stone that so

The Science and Commerce of Whisky
By Ian Buxton and Paul S. Hughes
© Buxton & Hughes, 2014
Published by the Royal Society of Chemistry, www.rsc.org

engaged the mediaeval mind and, from the standpoint of the Aristotelian view, represented an entirely logical pursuit.

Even today, an echo may be found in writing about distilling. Primo Levi, chemist, writes[1] in *The Periodic Table:*[†] "*Distilling is beautiful. First of all, because it is a slow, philosophic, and silent occupation, which keeps you busy but gives you time to think of other things, somewhat like riding a bike. Then, because it involves a metamorphosis from liquid to vapour (invisible), and from this once again to liquid; but in this double journey, up and down, purity is attained, an ambiguous and fascinating condition, which starts with chemistry and goes very far. And finally, when you set about distilling, you acquire the consciousness of repeating a ritual consecrated by the centuries, almost a religious act, in which from imperfect material you obtain the essence, the spirit, and in the first place alcohol, which gladdens the spirit and warms the heart.*"

The deliberate ambiguity of this passage, its overt reference to religion and its almost mystical tone – understandable, since Levi's skill as a chemist saved him from the forced labour gangs of Auschwitz – would, I suggest, be immediately familiar to the alchemist, despite proceeding from the formality and rigorous training of a modern scientist.

The inter-mingling of mystical symbolism, scientific practice and religious belief, so impenetrable to our contemporary mind-set with its emphasis on the rational and the material, leads us naturally to the monastery and the work of Franciscan friars, such as John of Rupescissa, Raymond Lull and the 13th century English philosopher and teacher Roger Bacon who, in 1267, attempted a synthesis of Aristotle's philosophy and science with contemporary theology, which he presented to his patron Pope Clement IV.

Basing much of their initial thinking and work on the Arabic writings of one Jabir ibn Hayyan (in Latin, Geber), who considered distillation the best way to separate nature into its component parts, they came to believe that the search for the fifth essence or 'water of life' would be through a series of distillations, often beginning with wine as the base. Silver and gold were also seen as incorruptible and therefore a suitable starting point for further transmutation. From this evolved the quest to change base metals

---

[†]Should any reader doubt the relevance of citing Primo Levi's The Periodic Table here, consider that in 2006 the Royal Institution considered it "the best science book ever."

into gold, the basis of much of the image of the alchemist and his search for the Philosopher's Stone in the popular imagination. However, as it is the medical application of distilling that led to distilled spirits as a beverage, the search for gold, however fascinating, is something of a sidebar to our story.

Perhaps the most famous description of distillation is given by Hieronymus Brunschwig.[2] *"Distilling is nothing other than purifying the gross from the subtle and the subtle from the gross...and the subtle spirit made more subtle so that it can better pierce and pass through the body...conveyed to the place most needful of health and comfort."*

The constant process of purification, as seen below in John French's series of linked alembics (Figure 1.1), was of critical importance to the alchemist (and it might be noted that this continues to be so to today's pharmacist or even distiller of vodka) and illustrations frequently show the 'pelican', a device for ensuring reflux and rectification. The pelican also carries symbolic meaning, referring to the legend that, in time of famine, the mother pelican would wound herself by striking her breast with her beak to feed her young with her blood to prevent starvation. By extension, early Christians adapted the pelican to symbolise Jesus as the Redeemer. The red blood of the pelican was also suggestive of a distillation process that had achieved the formation of the 'Red Tincture' and was close to reaching its ultimate stage of final transmutation (symbolised by the rebirth of the phoenix from the fire of distillation).

Figure 1.1 Apparatus for redistillation from John French's Art of Distillation (1651).

Thus, it is no coincidence that the first written reference to whisky in Scotland (from the Exchequer Rolls of 1494, discussed further in chapter 2) directs us to the monastery at Lindores Abbey or that, in 1510, Dom Bernado Vincelli first prepared the liqueur that today we know as Bénédictine in the Abbey of Fécamp in Normandy. To this day, the bottle also features the Latin motto of the Bénédictine order – *"Deo Optimo Maximo"* meaning "to God, the good, the great", as well as the coat of arms of Fécamp Abbey. Its counterpart, Chartreuse, dates originally from 1605 though it was not until 1737 that the liqueur was released to the world in a form resembling today's version.

While those traditions continue, little today remains of Lindores but excavations at Pontefract and Selborne Priories have revealed fragments of 15th century alembics. These were made of glass or pottery, and thus intrinsically fragile, or possibly of pewter, which would have been re-used in another vessel once redundant for distilling. Very little thus survives of this early technology but based on the Pontefract and Selborne excavations and earlier work, archaeologists have suggested the conjectural evolution of the still (see Figure 1.2).

As noted by Greenaway:[3] *"medieval Europe gradually developed a taste for distilled alcohol, at first generally in the form of liqueurs sweetened and flavoured by infusing leaves, etc., or by distillation from a mixture. More efficient distillation gave stronger distillates and eventually produced the aquavits and the brandy-wines, which are very strong indeed if drunk exactly as distilled. The abuse of these drinks is a part of social history. There was no dividing line between regimen and pharmacy in early times. The new strong drinks gave a feeling of warmth and well-being, which led to their being prescribed from the 14th century onwards for conditions producing feelings of chill and debility...so it is no accident that liqueurs and monasteries are commonly linked."*

And it is in a monastery that the story of Scotch whisky at least is said to start, which is discussed in detail in chapter 2.

1.2   IRELAND

The history of Irish distilling is a long, tangled and unfortunate one, containing a salutary warning against complacency, yet with a happy ending.

Most histories of whisky place its first arrival in Europe in Ireland and then Scotland, with Irish Christian monks having

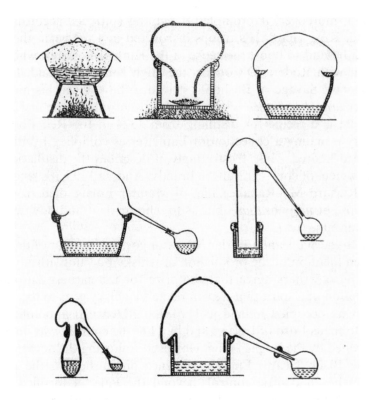

Figure 1.2   The conjectural evolution of the still.
From F. Greenaway *et al.*, (ref. 3) and reproduced in F. Sherwood
Taylor, *Annals of Science*, 1941–1947, vol. V.

obtained the secrets of distilling from the Moors in Spain, which
they then brought home. Though the principle of distillation was
known to the Arabs, the process by which it was transferred cannot
be documented or even dated with any certainty and, in any
event, it is far from clear that early distillation was used for the
production of beverage alcohol, being more disputed in making
medicines or perfumes.

It is not disputer that Irish monks travelled widely and gathered
learning. The monasteries they created were religious centres but
also places of great culture. Many maintained hospitals to treat the
local population and knowledge of the 'water of life' would have
been greatly prized. But, as well as a medicine, its use as a re-
storative would have been soon appreciated, especially in Ireland's
damp climate.

And, moreover, it promoted a fighting spirit as, according to legend, King Henry II's troops discovered in 1170 when they invaded Ireland to boost the cause of the King of Leinster, who was at war with Roderic O'Connor, the High King of Ireland. Later, Sir Robert Savage of Bushmills is said to have given his men "*a mighty draught of uisce beathe*" as they went into battle in 1276.

There is a recipe for distilling from 1324 in the Red Book of Ossory – mainly a collection of Latin verses complied by Bishop Richard Ledred – but, frustratingly, it describes the distillation of wine which, of course, results in brandy. And, in 1405, it's recorded that Richard MacRanall, Chief of Mainter Eolais, died from an overdose of *uisge beatha* – but as to who made it and how we are left wanting.

So to the Scots and to Friar John Cor goes the honour of the first written mention of the production of whisky but, notwithstanding that, most writers agree that the prize for the earliest European distillation of a spirit that we can relate to whisky goes to the Irish.

Having occupied Ireland in 1170, King Henry II appointed his son, John, as Lord of Ireland and by 1177 the country was directly controlled by the English king. However, following the devastation of the Black Death, English influence diminished to the point where they had little control 'beyond the Pale', a fortified area round Dublin, central English authority having withered away to little or nothing in the country.

And there, outside the reach of the English tax collectors, distilling quietly flourished. Just as in Scotland, distilling was an everyday and unremarked fact of the rural economy and the life of any substantial house. Attempts to raise tax were frustrated and this situation was to continue until King Henry VIII's invasion and subsequent domination of Ireland from 1536. Henry VIII then took a more practical view of the matter, attempting to raise tax and introducing a limit of one licensed distiller in a borough with substantial fines for anyone caught producing illicitly. Following his dissolution of the monasteries, there was a further dissemination of the skills of distilling and the relevant technology – in 1541 his successor, Queen Elizabeth I, is said to have received (and more importantly enjoyed) a Bishop's gift of a cask of whiskey. The story is widely repeated in whisky histories; however, as she was 8 years old at the time, perhaps it was enjoyed more by her courtiers than the Queen herself!

By 1577, Raphael Holinshed praised *aqua vitae* in his *History of Ireland* by remarking: *"truly it is a sovereign liquor if it be orderly taken"*. During the long Anglo – Spanish war, a failed Spanish invasion in 1601 soon led to a period of complete English dominance in Ireland and, in 1608, a system of licensing was introduced under King James I of England and VI of Scotland.

The indefatigable Elizabethan traveller, Fynes Moryson, who was employed in Ireland around 1600 later wrote[4] that *"the Irish aqua vitae, vulgarly called usquebagh, is held the best in the world of that kind, which is made also in England, but nothing so good as that which is brought out of Ireland. And the usquebagh is preferred before our aqua vitae, because the mingling of raisins, fennel seed, and other things, mitigating the heat, and making the taste pleasant, makes it less inflame, and yet refresh the weak stomach with moderate heat and good relish"*.

The quality and appeal of whiskey had not escaped the sharp eyes of the English administrators in Ireland's ruling class. Bishop George Montgomery wrote to his sister in November 1607 with a seasonal gift of Irish whiskey:

"I am appointed a Commissioner for the plotting and devvyding of the contreye (i.e. Ulster), which I feare mee will keep mee here this Christmas agaynst my will; and agaynst my will it shal be indeed yf I eat not som of my coson's Beaumont's Christmas pyes, and so tell her I praye you. I hope my sister and she have receaved the water I sent them in a little runlet of a pottle, a quart for a peece".

Famously, Sir Thomas Phillips paid 13 shillings 4 dimes for the right to distil for Coleraine and the Route. To this day, Bushmills have the date of 1608 embossed on their bottles, though this is a generous view of the distillery's foundation. In any event, the privilege was cancelled in 1620 following complaints of abuse and favouritism.

A tax of 4 pence per gallon was introduced on Christmas Day 1661 and, under this Act, Excise Commissioners were appointed for the first time with authority to appoint gaugers and searchers. Their effectiveness was limited, however, as the rule of law in Ireland was patchy at best, especially in more remote country areas and with few records it is hard to reliably estimate the extent to which distilling was carried out at this time.

The situation changed with new laws in 1717, 1719, 1731, 1741, 1751, 1759 and 1761 – the amount of legislation indicating the extent to which government was seeking to regulate and control the distilling industry. In particular, the duties and powers of the gauger were increased and distillery operations more closely regulated than hitherto.

However, according to E. B. McGuire,[5] imported rum was more popular than whiskey with over 2 million gallons imported in 1771, remaining dominant until the end of the 18th century, when Dublin began to emerge as a significant distilling centre. It should, of course, be remembered that as in Scotland distilling could be suspended by law at times of poor harvest or, in Ireland, actual famine. This happened in 1758–1759, for example, and again in 1765–1766.

That Irish whiskey enjoyed a high reputation in England is vividly demonstrated by Dr Samuel Johnson's Dictionary (1755) in which he defines usqueba'ugh as shown in Figure 1.3.

The reference to the spirit "*being drawn on aromaticks*" is interesting, suggesting as it does a product closer in character to today's gin with its botanicals than to whiskey. It is, however, the case that much 18th century usquebaugh appears to have been flavoured in some way – not just in Ireland – presumably to disguise the harsh nature of the unaged spirit.

Tax increases continued to assist the illicit distiller and increase the appeal of poteen. It is claimed that there were 2000 or more illicit stills in operation towards the end of the 18th century, though by 1796, 214 were licensed and there was a growing trend towards more substantial operations. Licenced distillers were also not averse to evading duty and there is considerable evidence that the same distillery might produce both duty paid and illicit whiskey, making production and revenue statistics highly unreliable.

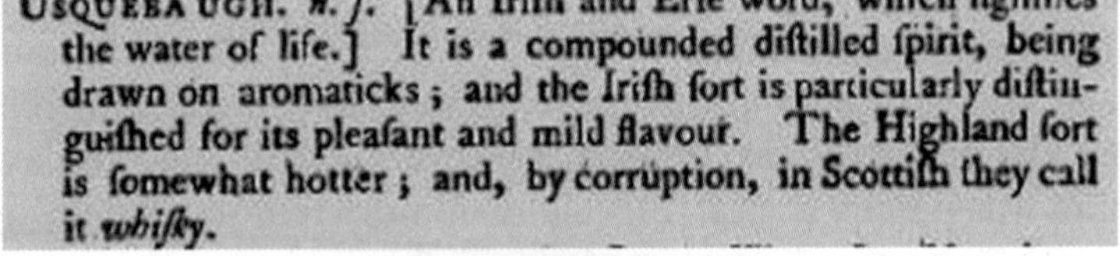

Usqueba'ugh. *n. f.* [An Irish and Erse word, which signifies the water of life.] It is a compounded distilled spirit, being drawn on aromaticks ; and the Irish sort is particularly distinguished for its pleasant and mild flavour. The Highland sort is somewhat hotter ; and, by corruption, in Scottish they call it *whisky.*

Figure 1.3   Dr. Samuel Johnson's 1755 definition.

There was extensive political agitation in Ireland during this period for free trade and to allow Irish goods access to the rest of the British Empire. Important legislation in 1780 started this process and the ascendancy of Irish whiskey in general and the Dublin distillers in particular began, though Cork was also established as a significant regional centre.

The four great Dublin distilleries began operating around this time, though the exact date of their establishment is sometimes unclear as the records are partial. For example, Jameson's quote their date of foundation as 1780 but the 1802 excise return identifies only two operating distilleries in Bow Street and Smithfield: Edmond Grange, then Dublin's pre-eminent distiller and John Stein, from the Scottish family of distillers of the same name. Stein and Edgar also had premises in Marrowbone Lane but, by 1802 that business was registered as Jameson and Stein and by 1822 it was known as William Jameson & Company (Figure 1.4).

Presumably, John Jameson must have acquired and expanded the Bow Street distilleries and, in 1810, he named the company John Jameson & Son. Eventually, four large Dublin distillers emerged: John Jameson & Son of Bow Street, John Power & Son of

Figure 1.4 Jameson's Bow Street distillery, Dublin, c.1878.

John's Lane, George Roe & Co. of Thomas Street, and Willam Jameson & Co. of Marrowbone Lane. They were to come to dominate the Irish trade.

While their exact history is hard to disentangle, and perhaps of little relevance at this distance in time, several factors emerge: these were businesses, employing large numbers of specialist staff where, increasingly, the owners and directors were distant from physical operations; they were dynastic in nature and they were, in their heyday, to prove immensely profitable.

One other name of particular interest that appears in the rank of Dublin distillers is that of Aeneas Coffey. A former excise man of some 25 years distinguished service, in which he was assaulted and severely injured by illicit distillers, engaged in a vigorous published debate with the Reverend E. Chichester on the 'oppressions and cruelties' of the Revenue's officers (whilst understanding the temptation and appeal of illicit distilling, Coffey took the side of the law), rose to the office of Inspector-General and proposed a number of technical innovations concerning still safes and revenue locks, he resigned in 1824 and began distilling at the Dock Distillery, Dodder Bank, Dublin. This passed to his son but closed by 1847.

His fame, however, rests on his development of the continuous, 'patent' or Coffey Still, which was radically to transform the world of whisky distilling. Coffey obtained his patent in February 1831 and was soon in the business of manufacturing stills. Around 1835, he relocated to London, where he continued operations until his death in 1852. The firm continues to this day under the name of John Dore & Co in Guildford, Surrey.

The Coffey Still took to their logical conclusion the very rapidly worked stills of the Scottish Lowland distillers, such as John Haig and Robert Stein, who had themselves patented a design for a continuous still in 1826. It took a few years for this technology to be widely adopted but, after the Gladstone's Spirit Act of 1860, which allowed blending in bonded warehouses before duty had to be paid, the Scots embraced the product of the continuous still with increasing enthusiasm. However, to their eventual cost, the Irish pot still industry took an entirely different view: a clash both of commercial interests and of cultures.

While history has preserved the reputation of the Dublin distillers, brief mention should be made of their rivals in Belfast and Londonderry which, until the closure of Dunville's (once the largest distillery in the United Kingdom) and United Distilleries in the late 1930s, represented a significant second force accounting for more than 90% of the distilling capacity of the north of Ireland. Today, only Bushmills remains, though plans to operate small craft distilleries in both Belfast and Campsie in Derry appear well advanced. Bushmills has in recent years been significantly expanded by its current owner, Diageo.

Irish whiskey grew rapidly from the 1820s, especially following the Distillery Act of 1823 and, according to Revenue statistics, by 1900 legally recorded production was approximately 14.5 million proof gallons from 30 distilleries.

Throughout the 19th century Irish whiskey enjoyed a markedly superior reputation compared to Scotch and dominated the home market. Still sizes had grown dramatically and this, combined with the practice of triple distillation, meant that the Irish product was smoother and more consistent than Scotch single malt.

This was due, in no small measure, to the size of their stills: Roe's Thomas Street distillery was then the largest pot-still distillery in the United Kingdom and the scale of the Dublin 'Big Four' should not be under-estimated. In his *Whisky Distilleries of the United Kingdom*,[6] Alfred Barnard gives details of their annual production in 1887:

- Jameson's Bow Street distillery: 1 million gallons (4.54m litres),
- Power's in John's Lane: 900 000 gallons (just over 4m litres),
- William Jameson's distillery in Marrowbone Lane: 900 000 gallons
- Roe's massive Thomas Street operation: nearly 2 million gallons (some 9m litres).

Contrast this with the scale of The Glenlivet at the same time: Barnard quotes this as *"nearly 200 000 gallons"* or less than one quarter of the smallest of the Dublin concerns. If any doubt remains about the status of the Dublin distilling industry at this time, consider that Barnard chose as the frontispiece of his book an engraving of the founder of John Power & Son. These were truly the giants of the 19th century whisky world.

However, to the consternation of the Irish trade, a practice grew up of shipping Scotch whisky to Ireland where, after a short period in warehouse, it could be re-exported as 'Irish' — a practice vehemently and vociferously opposed by the 'Big Four' in a series of pamphlets and campaigning around 1878 – 1879, culminating in the issue of a book by them: *Truths about Whisky.*[7] Questions were asked by Irish MPs in the House of Commons.

Truths about Whisky also rails against adoption of the product of the continuous still – or 'silent spirit' and 'sham whiskey', which refer to the Dubliners styled grain and blended whiskey, respectively. But be that as it may, political manoeuvring was of little weight compared to the reaction of the market. The energetic adoption of blending by the Scotch whisky industry saw the foundation and spectacularly rapid growth of some great firms: Dewar's, Walkers of Kilmarnock, Buchanan's, Haig & Haig and many other well-known concerns all prospered mightily at this time at the expense of the Irish distillers.

Several factors combined to aid the Scots: the innate shrewdness of their leading firms, the fashionability of all things Scottish, the spread of the British Empire and the arrival of phylloxera in European vineyards all played their part. But there is no doubt that they were greatly aided by an almost wilful refusal by the Irish firms to accept that blending had arrived and was being taken up enthusiastically by the consumer.

Like it or not, 'silent spirit' was here to stay; something which was subsequently confirmed by the Royal Commission on Whiskey, set up to determine the 'What is Whisky?' question, and which reported in favour of blending in 1909.

Other factors combined to undermine the strength of the Irish position, not least an over-reliance on one market, the USA, where over 400 brands of Irish whiskey were on sale by the late 19th century. The arrival of National Prohibition in the USA in January 1920 was followed by the Irish War of Independence with its unfavourable impact on sentiment in the English market. The eventual secession of the Irish Free State from the UK in 1922 triggered retribution in the form of high tariff barriers effectively blocking Irish whiskey exports to the remainder of the British Empire. A decision in 1926 by the Free State to increase the minimum age of Irish whiskey to 5 years, though doubtlessly well intentioned, placed the Irish industry at a further competitive disadvantage and punitive duty increases completed a dolorous picture.

When prohibition ended in 1933, the Scots proved to be in the stronger position, able to exploit the opportunity more energetically and with greater commercial acumen. An exhausted Irish industry appeared to have lost its competitive spirit. Many small distilleries had simply closed their doors, never to re-open. Avoniel, Belfast (closed 1929);[‡] Connswater, Belfast (closed 1929); Bandon, Co., Cork (closed 1929); Glen, Kilnap, Co., Cork (closed 1925); North Mall, Cork (closed 1925); Phoenix Park, Dublin (closed 1921) and Monasterevan, Co., Kildare (closed 1921) stand as examples of a depressing roll call of failure.

Post-war decline was even more marked – in fact, some commentators go so far as to suggest that Irish whiskey survived principally on the back of Irish coffee, popularised as an after-dinner drink. Whatever the truth of that, by 1966, only five distilleries remained working in the Republic and together they came to form the Irish Distillers Company (later, Irish Distillers Group). Today, after surrendering ownership to Seagrams, it is part of Pernod Ricard of France. The Dublin sites were finally closed, with only the Old Jameson distillery on Bow Street remaining as a visitor attraction.

The Midleton distillery at Cork was closed and re-modelled as a visitor centre and heritage site, while a brand new, state-of-the-art and multi-purpose distillery was built adjacent to it in 1975; although the individual brand identities are still maintained. Capable of producing a remarkable range of whiskies (and other spirits), the new Midleton distillery has been expanded on several occasions and as recently as 2013 at a cost in excess of €100m.

After several changes of ownership, Bushmills in the north is today owned by Diageo and has also received significant investment.

A recent domestic entrant to what is today a newly resurgent industry is Cooley, established in 1987 in the old Ceimici Teo distillery in County Louth, where vodka was distilled from potato by-products. John Teeling, an Irish entrepreneur, recognised an opportunity for a small Irish distiller and renamed the Dundalk-based distillery 'Cooley', making whiskey there from 1989 and going on to acquire other famous brands and distilleries which had

---

[‡]Avoniel enjoys a curious distinction as being the sole distillery to refuse access to Alfred Barnard who, understandably aggrieved, wrote: "*the proprietor stands conspicuous as being unwilling to allow an inspection of his works – for what reason we are unable to explain*". The proprietor's son did, however, provide Barnard with some scanty statistics.

been mothballed, including Tyrconnell and Kilbeggan. The Kilbeggan distillery, dating from 1757 and claiming to be the oldest distillery in the world, was re-opened in 2007 and a small stillhouse installed. In December 2011, the company was acquired by Beam Inc. of the USA. John Teeling and his eldest son, Jack, subsequently left the business to establish the Teeling Whiskey Co., aiming to *"revive the independent spirit of Ireland"*.

William Grant & Sons of Dufftown, Scotland acquired the Tullamore DEW brand from Irish Distillers in July 2010 and are currently developing a new, state-of-the-art pot still whiskey and malt whiskey distillery there at a cost of €35m.

Finally, a number of small boutique operations have been proposed and may be underway before this book reaches print, including sites in Dingle, Belfast, Carlow, Horse Island and elsewhere. Plans for these have been publicly revealed and there are virtually certain to be other projects that currently remain confidential.

Irish whiskey is, therefore, in as good a condition as it has been for more than a century. The category has been developed by shrewd, single-minded marketing led by Irish Distillers and Diageo; by growing consumer interest in products seen as authentic and with a distinct provenance and by-product delivery that is both easily accessible (yet, at its best, complex and intriguing) and distinctly different from Scotch whisky. It is a remarkable story of an entire industry brought back from the brink of oblivion.

1.3 CANADA AND THE UNITED STATES OF AMERICA

1.3.1 Canada

"Whereas, for the better support of the Government of this Province, it is expedient to increase the Revenues thereof..."

As is so often the case, one of the earliest mentions of whisky comes through tax records – here, it was the 1794 Act *"to lay and collect a duty upon stills"*, which was helpfully passed by the British Parliament into the Laws of the Province of Upper Canada[§] – and which remained the country's principal source of revenue for the

---

[§]What was originally Quebec Province was split into Upper Canada (Ontario) and Lower Canada (Quebec) by George III in June 1791. They were unified in 1840.

next half century. Given that America's Whiskey Rebellion had just been suppressed by George Washington's fledgling government, perhaps this was a bold move, but the Canadian trade acquiesced and within seven years 51 licences had been issued to a population of less than 15 000 people. It is tempting to see in this an early indication of the stereotypical law-abiding Canadian personality but it is also certain that there were many more stills operating at a domestic level, or illegally.

The still tax was set at 1 shilling and 3 dimes per gallon but, importantly, a minimum 10 gallon limit was set on the still size, with a draconian £10 penalty for infringement. This effectively outlawed small scale distilling.

So, who were these early distillers? We know that the first recorded distillery was established in 1769 in Quebec to make rum, which was popular in the Atlantic provinces and all the way down the eastern seaboard of North America. According to Lorraine Brown,[8] Lower Canada was also a whisky-making province but much of the output was destined for Europe. Upper Canada was to be the birthplace of Canadian whisky as we know it today.

With settlers from most of Europe, a variety of distilling traditions arrived in Canada. It might be expected that, as in the USA, whisky was established by Irish and Scottish settlers, who certainly had a huge part to play in building the country, but it was the English who were behind the early operations. Some commentaries have suggested that United Empire loyalists fleeing from the USA after 1776 brought distilling skills with them but, persuasive as this theory may be, no evidence remains.

The first recorded operation of scale was that of John Molson, a young Lincolnshire farmer who had emigrated to Montreal to set up as a brewer in 1783. In 1799 (or 1801, depending on which version of the story you prefer) he bought a still from the rum distillers, McBeath & Sheppard, and began making small quantities of whisky. His son, Thomas, took operations onto a higher level, travelling to Scotland and England in 1815–1816 to visit breweries and distilleries and learn his trade and, from 1821, he started to distil on a larger scale. Soon, he was shipping New Make spirit to London merchants, presumably for rectification in the London gin market. Again according to Brown, there was a substantial traffic in Canadian whisky to England until 1846 when it appears that Irish and Scottish distillers captured the market.

Molson was not the only distillery in Lower Canada. By 1827, thirty one concerns are recorded but, within four years, the total had increased to seventy; however, there were all minnows compared to the growing Molson operation which, in 1831, opened a second distillery at Kingston. By the 1840s around 200 distilleries were recorded in Canada as a whole.

Other distillers in Upper Canada included Richard Cartwright and James Morton. Morton's new distillery in Kingston was described as "*the handsomest and best finished establishment of the kind in British America or the United States*". After expanding into property, shipping, lumber milling and a locomotive works, the unfortunate Morton was bankrupted by the recession of 1857. Owing his bankers more than $250 000, he died in 1864 in a bed he had loaned back from them after they purchased all his furniture at auction. Budding entrepreneurs take note! The distillery eventually closed in 1900.

Although Thomas Molson seems to have been a talented distiller and the concern a substantial one, the family eventually decided to concentrate on brewing and, by 1867, they had closed their Montreal distillery. Just a few years earlier (1863) they had produced 336 000 gallons. By contrast, Dow & Company are recorded as distilling 204 000 gallons and Laprairie just 1207. However, as they were changing direction, other English migrants were busy developing their whisky business.

James Worts arrived in Toronto in 1831 and was shortly followed by his brother-in-law, William Gooderham. Back home they were millers and enthusiastic users of wind power. So, at the mouth of the River Don they erected a six-storey windmill and opened a grist mill. James died in 1834 but three years later, on November 3rd, the firm began distilling with James' son (also called James), who was appointed a partner in 1845. As such, Gooderham & Worts was established.

Production was in an early column still, made of wood and using cobblestones in the columns where today a metal plate or tray would be employed. The spirit was filtered through charcoal but, by 1862, was undergoing a second distillation after which the practice of filtration was abandoned.

In 1861, they built a splendid new distillery, the largest in Canada and capable of producing 2.5 million gallons annually. Two copper stills, each of 1500 gallons capacity produced 'Common

Whisky', 'Toddy' and 'Old Rye' — the latter two varieties apostrophised by the *Toronto Globe* as *"without question the very best and purest manufactured"*. Whisky was shipped to Lower Canada and, it is said, in large quantities to Liverpool and London.

Waste products from the distillery fed 400 cows, which produced milk for the city and 1000 head of beef cattle were slaughtered annually. Some 150 employees were engaged in the whole enterprise. Considering that this Gooderham & Worts distillery was some seven and a half times the size of their rival Molson's, the latter's decision to concentrate their efforts elsewhere is understandable.

Henry Corby, a baker from London, began milling but began distilling in 1859 and by the time of his death had established a flourishing business, which continued to grow under the direction of his son (also called Henry, but known to all as Harry). In 1905, however, the family sold their interests and two years later a fire destroyed the original distillery, which was then replaced by a new, larger one to serve a growing market in the USA and in Ontario, which had restricted alcohol sales in 1916.

At the end of World War I, during which time whisky distilling had been suspended by the government, the company was sold to the glamorously named Canadian Industrial Alcohol Company Ltd who did, however, acquire various wine and spirit import businesses and a firm of Scotch whisky blenders. The parent company also owned the J. P. Wiser distillery.

Much cross-border business was done during the period of US Prohibition and, although a key executive defected to re-start the Gooderham & Worts distillery in Toronto, sales boomed. That executive was Harry Hatch and, in 1932, he was able to purchase a 51% controlling interest in Corby's from Canadian Industrial Alcohol as Hiram Walker–Gooderham & Worts Ltd.

During World War II production was again suspended in favour of industrial alcohol but prosperity returned in the post-war period and the company built an additional rum and liqueur business with interests in the Lamb's Rum and Tia Maria brands. In 1978, in partnership with DeKuyper of Holland, Corby purchased Meagher's Distillery Ltd of Montréal and its subsidiary, the William Mara Company of Toronto, bringing representation of such international brands as Beefeater gin, along with an enhanced wine portfolio. The Corbyville distillery was expanded.

In 1987, Allied Lyons PLC successfully acquired a majority of Hiram Walker–Gooderham & Worts Ltd shares, thereby becoming Corby's majority shareholder. The Robert MacNish Scotch Company was sold and McGuinness Distilling Co. Ltd Was purchased with brands such as Polar Ice. In 1989, the Corbyville distillery was closed after 132 years and all maturing whisky inventory was transferred to the Hiram Walker facility in Windsor, Ontario. There was a short-lived foray into craft brewing but, in 1994, Allied Lyons PLC acquired Casa Pedro Domecq to become Allied Domecq PLC which, at the time, was the world's third largest spirits company.

However, Allied Domecq PLC was itself acquired by Pernod Ricard in 2005 and, at the time of writing, Pernod Ricard are the majority owners of the Corby company, as well as the proprietors of Hiram Walker & Sons, who today bottle and blend the majority of Corby's owned and agency brands.

One of the most respected names in Canadian distilling, according to Hiram Walker & Sons, is the only 'grain to glass' operation in Ontario and boasts the largest distillery capacity in North America with 37 fermenters. The manufacturing process distils 180 000 litres of alcohol every 24 hours and operates 24 hours a day, five days a week to produce a variety of products, including vodka, rum and Canadian whisky.

Again, this company's roots go back to an immigrant – in this case one Hiram Walker, an American grocer and vinegar distiller who moved from Detroit to Ontario and established his whisky distillery in 1858. A dynamic entrepreneur, he purchased a significant land holding near Windsor and established facilities for his workforce in what became known as Walkerville.

The company was spectacularly successful from the start, with a flourishing cross-border trade that boomed during the American Civil War. Surviving the recessionary economy of the 1870s, Hiram Walker launched Canadian Club in 1884, a brand which swiftly gained favour in the USA where it was widely imitated by a number of copycat brands and outright fakes. Considerable controversy ensued and a lively debate on fusel oil in whiskey eventually led to a US Government enquiry.[¶] This vindicated Canadian

---

[¶] The campaigning pamphlet "*A Plot Against the People*", issued by Hiram Walker in 1911 as they lobbied against the Kentucky whiskey interests, is a minor classic of its kind.

Club, which went on to further success. Today, the brand is owned by Beam Inc. of the USA.

In 1926, Hiram Walker's grandsons sold the business to Harry Hatch's Gooderham & Worts company and, in 1933, they built what was at that time the world's largest distillery in Peoria, Illinois − then a major brewing and distilling centre. It closed in 1981.

Walker is remembered for a number of key innovations in Canadian distilling: the first multi-column distillation; arguably the first Canadian blend; the first temperature controlled warehouses; one of the first creators of brand names; the first to employ travelling salesmen and the first North American producer to receive a Royal Warrant from the British Crown (for Canadian Club, 1898). Later, in the 1940s, the firm introduced gift wrapped packaging, a radical marketing idea at the time, quickly adopted by competitors.

The original Walkerville distillery was renamed Wiser's following its acquisition by Pernod Ricard but remains notable for being the only Canadian distillery to use malted rye, a specific spring variety which must be purchased a year in advance. It also remains Canada's oldest distillery.

The Wiser name dates back to 1857 when a Prescott, Ontario businessman, called James Averell, employed J. P. Wiser and had him manage the Payne's distillery in the town, which he had just bought. Under Wiser's leadership, it grew in size and reputation and, within five years, he was able to buy the business.

Shortly afterwards, in 1864, the distillery was badly damaged by fire but Wiser rebuilt and expanded it − actions he was to repeat in 1887 following another fire. Wiser himself stressed quality and the importance of time in whisky making. Following his death in 1911, his sons took over the business but sold it in 1927 to Corby Distilleries. The Prescott distillery was closed in 1932 and today the Wiser brands are distilled in Walkerville.

The final great name in Canadian distilling history is that of Seagram's. The remarkable story of that company begins with Joseph Emm Seagram, from a family of English immigrants, who by 1883 had acquired the sole interest in a milling and distilling business in Waterloo. He expanded this and, in 1887, launched Seagram's 83, a four-year-old sherry cask-matured whisky.

By 1911, his sons had joined the business and, in March 1917, they launched Seagram's VO, destined to become their best-selling brand. In 1928, following some years of difficult trading and a public offering of the Waterloo distillery, the firm merged with the Distillers Corporation of Montreal, controlled by Samuel Bronfman.

The Bronfman family had arrived in Canada in 1889 as escapees from the persecution of the Jews in Russian Rumania. Davin de Kergommeaux says of Samuel Bronfman:[9] *"his constant self-aggrandizement, his need to hobnob with the elite, and his desire for a royal warrant if not a knighthood, tell of a man with a desperate need of validation by others. He could be foul-mouthed, cruel, and ruthless, yet he inspired loyalty, respect and perhaps even love in those who knew him well."*

The Distillers Corporation prospered during the Prohibition era, yet 'Mr Sam', as he became known, also appreciated the long-term nature of the whisky business and astutely developed the business with an emphasis on quality and brand building. Competitors, such as the British Columbia Distillery Co. of Vancouver, were bought out and a number of distilleries were acquired in the USA and elsewhere. A major new distillery had been built in Gimli, Manitoba in 1968. Thus, by 1970, the company was a global player operating 39 distilleries around the world and owning major Scotch whisky brands, such as The Glenlivet and Chivas Regal.

Samuel Bronfman, who had given lavishly to Jewish charities, died in 1971 and the business was inherited by his son, Edgar, who further expanded the business. However, the decline in the whisky market of the 1980s forced the company to retrench and the historic Waterloo distillery was finally closed in November 1992.

In 1994, Edgar Bronfman Jr took over as Chief Executive Officer. At that time, a shareholding in the Dupont Corporation was the company's largest source of income. In what many commentators regarded as a misguided move, Bronfman sold this stake for some $9 billion in order to fund acquisitions in the entertainment industry, which proved a short-lived and not entirely successful diversification. The drinks industry interests were acquired by Pernod Ricard who sold the Seagram name and brands to Diageo. Distilling continues at Gimli and Valleyfield in Quebec.

Today, other Canadian distilling operations include Alberta Distillers, Black Velvet and Highwood in Alberta – all three of which are largely anonymous or unknown outside of Canada – Canadian Mist (owned by Brown – Forman of the USA),

Valleyfield and two smaller operations: Glenora in Nova Scotia and Kittling Ridge, home to the Forty Creek boutique brand.

The history of Canadian whisky is complex and, at times, confusing. Canadian whisky is not – in general – well-known or fully appreciated outside of North America but it is a rich and fascinating history and a product deserving of greater attention. As De Kergommeaux concludes: *"the whisky world is waking up to Canadian whisky and declaring some of it as among the best whiskies in the world."*

1.3.2 United States of America

Before whiskey there was rum. And before rum the early settlers drank applejack, or apple brandy. Indeed, what is claimed to be *"America's Oldest Native Distillery"* Laird & Company traces itself back to the 1698 arrival from Scotland of William Laird, who allegedly began to use his skills as a distiller to process the abundant apple crops of New Jersey. You can still purchase Laird's applejack to this day – a tangible link to the first European colonists.

Rum was widely distilled in New England and actually formed part of the disreputable 'Triangular Trade', being exchanged for slaves in Africa, which were traded for molasses in the West Indies, which was in turn distilled in New England. As early as 1733 the British Crown attempted to tax molasses and rum produced in America, contributing to the colonists' feelings of discrimination, which led eventually to the American War of Independence (1775–1783). By the late 1790s, however, fashions had changed and whiskey began to displace rum, a trend that was accelerated by the 1808 ban on importing slaves from Africa, which dealt a fatal blow to the rum/slaves/molasses business.

The Treaty of Paris, ending the war with the British, was signed by Congress in 1784 marking the end of hostilities. By April 1789, George Washington had been elected the first President of the new United States and his name subsequently appears in American distilling history – both as a significant distiller (albeit with a Scottish distillery manager) and for his part in the Whiskey Rebellion of 1791.

Washington's distillery at his estate at historic Mount Vernon has been extensively documented and recently recreated (Figure 1.5). It is, however, not at all typical of the thousands of distilleries

Figure 1.5 Working in the recreation of George Washington's Mount Vernon
distillery. Courtesy: George Washington's Mount Vernon.

that were operating in the USA by 1800 as, for the time, it was a
substantial enterprise with five stills. First established in 1797, by
1799 (the year of Washington's death) it recorded an output of
nearly 11 000 gallons.

But prior to becoming a distillery owner, Washington had already secured his place in American distilling history. Scots
and Irish immigrants had settled in the western parts of Pennsylvania, Maryland, Virginia and the Carolinas and German settlers in Pennsylvania. Both groups found that rye grew well
(as opposed to barley, which did not) and so began making rye
whiskey. Meanwhile, in Kentucky County, especially around
the settlement of Bardstown, so-called 'corn patch' settlers and
those who followed began to use the native American corn in their
stills.

By 1791, the newly independent United States, with Washington
as the first President, needed money. Treasury Secretary Alexander
Hamilton proposed a tax on distilling, arguing that this was a
luxury product. In addition, some social reformers saw it as a tax
on sin that would lessen the pernicious effects of alcohol on society.
The embattled farmers of western Pennsylvania did not agree,

viewing this as taxation without representation, a cause over which many of them had all too recently fought the British. In addition, they faced logistical difficulties in paying the tax and its structure favoured the larger commercial distillers in the East.

A tax collector's house was burnt down; others were tarred and feathered and a simmering resentment of the tax and defiance of central Government continued until May 1794 when writs were issued to distillers who had not paid the tax. This required them to travel to Philadelphia to appear in federal court. For farmers on the western frontier, such a journey was expensive, time-consuming and not without dangers and soon the bitterness in the west gave way to an armed insurrection and open rebellion.

Having reduced and simplified the tax, but determined to enforce the authority of central Government, Washington mobilised an army of some 13 000 troops[||] and, in September and October 1794, marched into western Pennsylvania – where the insurrection promptly collapsed. Many of its leaders fled further west; some were captured and put on trial for treason. Two men were condemned to death, but received Presidential pardons; Washington's reputation and political capital and the authority of the Federal Government were both greatly increased by what was seen as his shrewd handling of the affair.

A grant of 60 acres of land in western Virginia was then available to any settler building a house and raising corn. Many of the disenchanted Pennsylvania farmers took advantage of this, bringing their skills and stills to what was to become Kentucky. Lighter in style than rye, a new whiskey was born.

Some controversy surrounds the varying claims to be the first distiller of bourbon, with Evan Williams, the Reverend Elijah Craig, Jacob Beam and others now recorded only in dusty ledgers and all suggested as the pioneer. Perhaps the best contender, however, is Dr James Crow (a Scotsman) who introduced the distinctive sour mash process,[**] essential to bourbon. In their comprehensive study of bourbon,[10] Gary and Mardee Haidin Regan concluded: "*for those who insist on having a name, we say James Crow 'invented' bourbon sometime between 1823 and 1845.*"

[||]Said to be a larger army than any that faced the British.
[**]In the sour mash process, some of an older batch of mash is retained and used to start fermentation in the following batch.

Around the same time, the 'Lincoln County Process' was invented, a system of filtration involving a deep bed of sugar-maple charcoal, through which the new whiskey was passed. The use of this lengthy filtration defines Tennessee whiskey as distinct from bourbon or rye. Today, the most notable brand using this method is Jack Daniel's, often mistakenly referred to as bourbon.

Writing in 1838, Samuel Morewood[11] records some 3594 stills in Pennsylvania; 2000 in Kentucky; 591 in New York; 560 Connecticut and so on until just 6 were noted in the Mississippi territory. Morewood's data covers the country as far west as the Illinois territory (an unknown number of stills producing just 10 200 gallons of spirit). However, his record is only a partial one: whereas the Pennsylvania stills produced just over 6.5m gallons of spirits, Virginia was recorded as making nearly 2.4m gallons and North Carolina close to 1.9m gallons; however, the number of stills was not recorded.

In total, Morewood records production of nearly 25.4m gallons. Based on his figures for known output from known stills, it is possible to calculate that there may have been a further 3564 undocumented stills, suggesting that around 11 300 stills were working at that time – excluding bootleg distillation of moonshine liquor, of course.

What this data demonstrates is that most distilling was still local in nature, largely due to the agrarian economy and the difficulties and cost of transportation across such a vast country. River transport was widely used. "*The immense number of navigable lakes and rivers which intersect this vast continent, affords great facility for the transportation of spirits,*" writes Morewood. "*In the course of eleven months, terminating on the 1st July, 1811, among other articles, 3768 barrels of whiskey were sent down the Shenandoah and Potomac rivers; while the spirits made at Brownsville, near Pittsburg, are in such repute that they are frequently sent to New Orleans, a distance of nearly 2000 miles. In the year 1822, 7500 barrels of whiskey, value 500 000 dollars...were sent from the western States for consumption.*" According to the 1820 Census, the population of the United States was 9 638 453 persons, of whom just over 1.5m were slaves. *Per capita* consumption, therefore, would have been impressively high by current standards but, as Sarah Hand Meacham notes:[12] "*alcohol was one of the few pleasures to be had in the early modern world.*"

However important the waterways were, the expansion of the railway network during the first half of the 19th century transformed the possibilities for national sales and marketing. However, during the same period, a burgeoning temperance movement began to gather strength, which would eventually play a pivotal role in the development of distilling in the USA. In considering the best-known impact of this movement, the imposition of National Prohibition from January 1920, it is important to recall that the temperance movement was deep-rooted and of long standing, with a number of northern States electing to go 'dry' from the 1850s.

Whiskey production was inevitably disrupted by the Civil War (1861–1865) and a number of distillers were forced out of business. The Federal tax on whiskey, reduced to zero in 1834, was re-introduced by Lincoln in 1862 to pay for the war and, in its conclusion, the industry went through a period of reorganisation and modernisation. A political scandal, known as the Whiskey Ring, which involved a tax fraud alleged to generate secret funds for the re-election of President Ulysses S. Grant, dragged on through the 1870s, creating rather more heat than light ; however, it besmirched whiskey's image.

Throughout the 19th century, as settlers pushed west, whiskey was employed as currency and, as it had been since the days of the French and British colonial rulers, to trade with the Native Americans, among whom it had disastrous effects. Unused to alcohol, and with no tradition of distilling, their habit was to share the bottle or barrel amongst friends and to continue drinking until it was all gone. The most infamous of the many trading posts where this disreputable trade was carried on was the notorious Fort Whoop-Up. Though located in Alberta, Canada it stands as a sad memorial to the exploitative behaviour of 'the white man' across all of North America.

A further scandal followed in the next decade. The Distillers' and Cattle Feeders' Trust of Peoria, Illinois[††] – known to one and all as the Whiskey Trust – attempted to gain monopoly control of the market by buying up and closing small distilleries. Some that refused were the victims of a mysterious arsonist! After various legal investigations, the Trust was forced out of existence but not before

---

[††]Illinois was, at that time, a major distilling centre. During the years of 1837 to 1919 Peoria housed 24 breweries and 73 distilleries but the activities of the Trust silenced most of them.

it had further damaged the image of the industry and forced a number of firms out of business.

The trend in any event was towards larger and larger distilleries, with the continuous still finding increasing favour. When we consider the data presented by Morewood, it is evident that rye whiskey was evidently the most successful variety during the first half of the 19th century and, indeed, it maintained that position well into the latter half of the century, until it was undone all but entirely by prohibition.

The practice of 'bonding' – keeping whiskey under Government supervision in a bonded warehouse, free of tax – was first introduced in 1868 and expanded in 1879, 1894 and again in 1897 with the Bottled in Bond Act, which assisted legitimate distillers in differentiating their products from that of the less scrupulous. Although glass bottles had been, expensively, available for some time prior to 1870 it was only then that George Garvin Brown, inspired by the pharmaceutical industry, began to sell his Old Forester brand entirely in sealed bottles.

There was considerable growth in the latter part of the 19th century, notwithstanding the energetic attacks of the temperance lobby, with many of the great names from Kentucky that would reappear after prohibition being established, along with continued buoyant sales of rye whiskies.

The USA's entry into World War I in April 1917 led to a ban on the distillation of all beverage alcohol to preserve grain for food. It was to prove a short-lived setback because the prohibitionist cause was gaining momentum.

Again, it cannot be stressed strongly enough that this was not a sudden or unexpected challenge. Travelling during 1892 in the USA on the first of his great international sales trips, whisky entrepreneur Thomas Dewar later recalled his experience of trying to buy whiskey in a prohibition state. The conductor on his train, unable to sell him a bottle, advised trying a store at the next stop and ordering 'cholera mixture' – *"I did, but to my great astonishment received a very familiar bottle which, although it was labelled on one side 'Cholera Mixture: a wine-glassful to be taken every two hours' had upon the other side the well-known label of a firm of Scotch whisky distillers, whose name modesty requires me to suppress!"*[13]

Like other distillers, Dewar was an energetic opponent of the effects of prohibition, observing in one of his famous aphorisms that *"if you forbid a man to do a thing, you will add the joy of piracy and the zest of smuggling to his life."* George Garvin Brown published a book of Biblical texts favouring alcohol and organisations such as The Association against the Prohibition Amendment ensured that the teetotal case did not go unanswered.

But the cause of the so-called Noble Experiment was not to be denied and, as state after state went dry a national ban became inevitable. The National Prohibition Act[‡‡] was passed by 287 votes to 100 in October 1919 and came into full force in January 1920.

Much has been written on this subject. Suffice to say that the law was only loosely enforced and soon fell into disrepute; that it proved a bonanza for distillers of Scotch and Canadian whiskies, who proved only too happy to supply their products with scant regard for the legal niceties and the backgrounds of their new customers; that organised crime soon infiltrated American society (and has subsequently proved hard to eradicate) and that alcohol consumption, especially of spirits appears to have increased during the Prohibition years.

All round, it was in fact an unmitigated disaster – especially for the estimated 15 000 victims afflicted by 'jake foot', a debilitating paralysis of the hands and feet brought on by drinking bootleg alcohol flavoured with ginger root. European visitors frequently commented on the ease with which drink could be obtained and the open defiance of the law: *"these Old Fashioneds will be the undoing of me"* wrote publisher George Blake[14] from a New York sales trip in November 1930 and, later: *"I am not, after six weeks of Old Fashioneds, quite trusting my own judgement."*[§§]

For the industry, it was, of course, a complete disaster. Although some distilleries were permitted to remain open to make 'medicinal alcohol', the vast majority closed for good. When prohibition was finally repealed in December 1933, there were insufficient stocks of domestically-distilled whiskey to meet demand and consumer tastes had adopted to Scotch and Canadian whiskies and to gin, a favoured product of the bootleg distiller. *"It was doggone scary"*,

---

[‡‡]Often known as the Volstead Act after its sponsor Andrew Volstead, a Republican Congressman from Minnesota.
[§§]He had made the error of drinking with the American author, Christopher Morley, noted for his enthusiastic consumption and great love of whiskey; brother to Frank, Blake's colleague at Faber & Faber.

according to Heaven Hill's Max Shapira,[¶¶] who went on to note that "*this was an industry that hadn't existed for 16 years; it was full of bootleggers; its product wasn't respected and there wasn't any of it anyway.*"

But no sooner had the beleaguered industry recovered some sort of footing than production of alcohol was diverted to the military effort of World War II and rum was very happy to fill the gap, with Scotch whisky favoured by returning GIs. Following the resumption of distilling the US industry concentrated on supplying cheap and cheerful spirits to the domestic market, only for more uncertainty to dog their efforts with a widespread belief that distilling would be suspended during the Korean War (1950–1953).

Although this, in fact, never happened, a concentration on the lower end of the market meant that, with vodka and white rum sales gathering pace and a public climate increasingly hostile to heavy consumption, by the mid-to-late 1970s, the sales of rye had all but dried up and bourbon sales were following it rapidly downhill.

Since then there has been a sustained revival, interspersed as is the way of these things by the occasional economic slowdown and major events, such as the Vietnam War; however, a revitalised industry has been able to ride out these setbacks and build a more robust industry, with a greater emphasis on premium products. Based initially on a trend to 'drink less but drink better' and assisted by the educational efforts of the Scotch malt whisky industry, which both pointed the US consumer to more aspirational products and gave US producers a business model to copy, there has been a revival in straight whiskeys.

At the centre of this move to higher quality products a few companies have been instrumental, most notably the Maker's Mark distillery (who were able to purchase their facility in Loretto, Kentucky in 1953 when the then owners sold up). Then independent and family owned, Maker's Mark is now part of Beam Inc., who also own the Jim Beam brand and a stable of small batch bourbons (Knob Creek, Basil Hayden's, Baker's and others), as well as the Old Grandad and Old Crow labels. Their Old Overholt rye refers to one of the founders of American distilling, Abraham

---

[¶¶]Quoted in *Spirits & Cocktails* by Dave Broom, Carlton Books: London, 1998.

Overholt, and the company also owns Canadian Club and a portfolio of Scotch, Irish and Spanish whiskies.

Amongst the other principal producers is Brown – Forman, proprietors of Jack Daniel's, Early Times, Old Forester and Woodford Reserve, a showpiece distillery and visitor centre. Although now a quoted company, the influence of the founding Brown family remains strong and they are actively engaged in its management.

Diageo are relatively poorly represented in this category given the overall size of the company, owning I. W. Harper and Bulleit bourbons and the George Dickel Tennessee whiskey. Analysts and commentators on the industry see Diageo as a likely buyer of Beam, should that company ever be a takeover target. Currently, most of Diageo's effort in this category is focussed on promoting the Bulleit brand.

Other major producers include Heaven Hill, Sazerac, Four Roses (owned by the Kirin Group of Japan), Wild Turkey (Gruppo Campari, Italy) and Lawrenceburg Distillers of Indiana, who produce for a wide range of brand owners. There are several hundred craft or boutique distillers (refer to chapter 7 for further commentary on this phenomenon).

In general terms, American whiskey has yet to achieve significant purchase in international markets with only a very small number of brands, most notably Jack Daniel's, Maker's Mark and Woodford Reserve, achieving significant sales outside the USA. Around 70% of bourbon sales are domestic, with Germany, the UK and Australia representing the principal export markets. The development of flavoured whiskies – chiefly honey variants – appears to have worked well for the US industry attracting new consumers into the category.

The future appears to be bright for both the traditional styles of bourbon and rye and for the more innovative products and the output of the craft sector. American whiskey would seem to have shaken off its past and is thus well placed for a sustainable long-term future.

1.4 JAPAN

Between 1603 and 1868, in what is known as the Edo period, Japan was effectively closed to Western influence. Apart from very limited access to an artificial island outside Nagasaki, all foreigners were banned from Japan and contact was severely restricted.

However, in 1853 a group of American warships under Commodore Matthew Perry arrived in Edo with the objective of opening Japan to foreign trade. Various treaties followed, but the process was not entirely peaceful; in 1864, British, French, Dutch and American warships bombarded Shimonoseki forcing more Japanese ports to open to foreigners. By 1868, the Tokugawa Shogunate had failed and the Meiji emperor was restored. This was a period of dramatic change in Japanese society.

Perry brought gifts for the Emperor and his senior Councillors. These included a steam engine and track, a copy of Audubon's *Birds*, a stove, a revolver, a telescope and whiskey (whether it was bourbon or rye is not recorded). One barrel was reserved for the Emperor, although he apparently never received it, and the various Councillors were allocated between one and five gallons apiece, along with quantities of cherry cordial.

Sake had been produced in Japan since around 710 AD but expanded rapidly during the Meiji period (1868–1912) when there may have been around 30 000 producers. According to Olive Checkland,[15] the manufacture of Western spirits (in particular, gin) was attempted in Japan from the start of the Meiji period. By 1912, when the Emperor died and co-incidentally taxes on imports were drastically increased, a recognizably westernised Japanese spirits industry began to develop.

Gisuke Konishi, a 'foreign drinks maker' in Osaka, was making and selling an ersatz 'whisky' from 1888 and Shinjiro Torii, his nephew, worked there as a young man. Torii subsequently founded a firm called Kotobukiya and launched a sweet wine product in 1907 – this was the basis of Suntory.

By 1919, he had launched 'Finest Liqueur Old Scotch Whisky', which carried a label stating it was bottled by the 'Torys Distillery', although there were no whisky distilleries in Japan at this time. Whisky labelling echoing or directly imitating Scotch whisky was to remain commonplace in Japan for many years afterwards, perhaps accounting for the story, surely apocryphal, of the Japanese town renamed 'Scotland' so that its whisky could be labelled 'Made in Scotland'. More than 40 years later, the official Scottish attitude towards Japanese whisky remained patronising and complacent.

Also in 1919, 25 year old Masataka Taketsuru, who had travelled from Japan in the previous year at the instigation of his then employer to learn how to make whisky in the Scottish manner, began his practical studies in distilling. Taketsuru came from a family of sake brewers and, in Scotland, spent some time studying chemistry at Glasgow University and the Royal Technical College (now Strathclyde University). He also worked for a week at Longmorn Distillery, having been disappointed in his hope of private tuition with J. A. Nettleton in Elgin. Although he had studied Nettleton's book on distilling and begun translating it into Japanese, Nettleton's proposed fees were far too high for him to consider and he was forced to turn to Longmorn, where General Manager J. R. Grant allowed him some practical experience. Today he might be considered an 'intern' as he worked without payment.

While Scotland presumably felt alien to Taketsuru, who would have stood out as a foreigner, he would have experienced curiosity but little overt hostility. Japan and Britain had signed the Anglo – Japanese Alliance in 1902, which was expanded in 1905 and 1911. During World War I, therefore, Japan fought on the British side and there were also cultural links with the UK.

On his return to Glasgow, where he was to marry, he studied the Coffey Still at Bo'ness Distillery and, with his new wife, spent five months in Campbeltown as an 'apprentice' at Hazelburn Distillery, where he studiously compiled a lengthy report into whisky production. In November 1920, he returned to Japan.

Masataka Taketsuru is generally regarded as the father of Japanese whisky but, as we shall see, his story is closely mingled with that of Suntory. However, his career initially began with disappointment as his initial employer and sponsor was forced to abandon plans to make whisky and, in 1922, Taketsuru resigned from the Settsu Shuzo company.

He then joined Shinjiro Torii at Kotobukiya (the company which later became Suntory). Some accounts place Torii amongst the group of friends and family who bade him farewell on his way to Scotland in 1918. Whatever the truth of that, Torii had prospered during World War I and was equally determined to produce a high quality whisky, so hired Taketsuru on generous terms with the aim of establishing a distillery on the Scottish model.

This was opened at Yamazaki in November 1924 and is generally considered the first Japanese whisky distillery – perhaps predictably, Suntory tend to credit Torii with the choice of the site, while other commentators give the more experienced Taketsuru the credit. Intriguingly, however, on the Nonjatta blog,[16] Chris Bunting suggests that there may be an earlier contender for Japan's first whisky, noting that Kotaro Miyazaki's Yamanashi distillery was marketing a 'K. M. Sweet Home Whisky' two years earlier.

According to Bunting some while later having changed its name to Daikoku Budoshu, the firm began distilling whisky and, after World War II, launched the Ocean brand, first using spirit sourced from Nikka, then distilling at a site called Shiojiri (now closed) in Nagano prefecture. Eventually, around 1955–1956, a move was made to Karuizawa distillery, which will feature later in the story of modern Japanese whisky.

There is also a suggestion that Eigashima's White Oak distillery may have marketed a 'Scotch whisky' around 1919 in what Bunting describes as *"those wild old days."* It is not clear, however, whether this product was made in Japan or imported. Today, the company offers blended whisky and a limited range of single malt.

Yamazaki's early production was unsatisfactory and Taketsuru returned to Scotland in 1925 for further technical studies at Hazelburn, shortly before it was closed. Taking the answers to his questions back to Japan, the first Suntory whisky Shirofuda (White Label) was launched. Today, Suntory's corporate website[17] describes this as *"true Scotch whisky, an accomplishment showcased in newspaper advertisements declaring that there would be 'no more imports needed'."*

However, having been effectively sidelined in his career by 1929, in 1934, Taketsuru resigned from Suntory to launch his own business, Dainipponkaju, later renamed Nikka Whisky. Initially, this was suggested by Checkland to produce apple juice and apple wine – out of respect for his erstwhile employer – but it seems clear that the site at Yoichi was always destined in Taketsuru's plans to produce whisky and stills were installed there in late 1935.

The company was initially loss-making but, as production switched to whisky, which was strongly favoured by the Imperial Japanese officer class and thus enjoyed a preferential allocation of fuel and raw materials, profits increased. Suntory's Kabukin brand, launched in 1937, also enjoyed military patronage and both companies benefited from being seen as vital to Japan's war effort – the country having by now switched allegiances to oppose the UK and the USA during World War II.

After the war, whisky began to assume extraordinary levels of popularity in Japan and the distilling industry there entered a period of prosperity, albeit with intense levels of competition. Initially drunk by the occupying forces, whisky became the drink of choice for 'salarymen', for business entertaining and for the large numbers of American troops stationed in Japan or later posted there for rest periods during the Korean and later the Vietnam war. Suntory finally launched their Old brand in 1950 after a 10 year delay. Tax changes in 1953 dramatically decreased the cost to the consumer of cheaper blends, further boosting consumption.

Increasing status began to be attached to gifting more and more prestigious whiskies in business life and specialist whisky bars began to thrive. Notable amongst these were Suntory's Torys bars – the company operated over 1500 such establishments at their peak of popularity.

As well as Suntory and Nikka, other Japanese distillers of the post-War period included the Takara, Toyo, Tokyo and Daikoku Budoshu distilling companies. Production was virtually entirely for home consumption with the majority of spirits being employed in blending. Bulk quantities of single malt whisky were imported from Scotland and used to enhance the quality of domestic spirits, business which became increasingly controversial in Scotland and often relying on brokers to fulfil orders. The brokers and distillers Stanley P. Morrison Ltd (later Morrison Bowmore) were at the front of this trade and the firm was eventually acquired by Suntory.

Nikka opened their second distillery at Miyagikyo (sometimes referred to as Sendai) in 1969 and Suntory followed with the giant Hakushu distillery in 1973, which was further expanded in 1981. In 1984, they launched Yamasaki 12 Year Old, the first Japanese single malt of any international significance.

The Japanese economy grew strongly in the 1980s and, with it, the domestic demand for whisky, both home produced and imported, reached a high point around 1982. However, the booming economy led to an asset price bubble and the subsequent crash, dating from sharp interest rate rises in late 1989, resulted in Japan's lost economic decade, which some critics maintain has continued to run now for over 20 years. The domestic industry was also protected by high tariffs, which were reduced in 1989, leading to difficulties for many producers as they were caught between cheaper foreign imports and the sudden fashionability of shochu.

As a result, production was cut back (analogous to the reaction to Scotland's infamous 'Whisky Loch') and a number of distilleries were mothballed or permanently closed, including Shinshu, Karuizawa, Hakushu West, Shirakawa, Kagoshima, Yamanshi and Hanyu.

Since the new millennium, however, Japanese whisky has again been in the ascendency, winning a number of major awards and, for the first time, achieving some international awareness and distribution, even if primarily in specialist whisky retailers and bars. Whilst this has been mainly driven by Japanese single malts, which have achieved a level of cult following and high prices,[‖] some blended whiskies (notably Hibiki) have made an impression outside their home market. Japanese producers are, however, somewhat hamstrung by the lack of older stock due to production cutbacks in the 1990s and this has had an adverse impact on their ability to compete fully in international markets (Figure 1.6).

Figure 1.6   A £10 000 bottle of single malt from the Karuizawa distillery. Cask #3603, distilled 1964. Courtesy of Number One Drinks Company Ltd.

[‖] See, for example, the 2013 release of a 1964 distillation from the closed Karuizawa distillery, which retailed at £10,000 per bottle. See Figure 1.6.

Outside of Japan, distilleries, such as Ben Nevis, Tomatin, Bowmore, Auchentoshan and Glen Garioch in Scotland and Four Roses in Kentucky, are in Japanese ownership, a further mark of the growing globalisation of the whisky industry.

Finally, though there was little or no tradition of artisanal or farmhouse distilling of whisky, recent years have seen the establishment of new small distilleries at Shinshu (reportedly using the original equipment) and Chichibu. This trend may well represent a further stage in the renaissance in Japanese whisky distilling, which it should be recalled, has a history of less than 100 years.

Its growth and development, notwithstanding well-documented setbacks, has been nothing less than remarkable. From ersatz whisky to consciously matching the best of Scottish practice, Japanese whisky has now developed its own distinct and highly-regarded style and personality and fully deserves its place on the world stage.

1.5 WORLD WHISKY

Chapter 5 contains a fuller review of developments in new 'world whisky' distillation. This has been a remarkable phenomenon, which has gathered pace in recent years and is continuing to grow rapidly, albeit from a tiny base.

The distillation of whisky has spread remarkably and now can be found on every continent barring Antartica (although it is not inconceivable that, in a hut somewhere, in the long polar nights, a still has been fired up...) In consequence, whisky has become a truly international spirit: as Aeneas MacDonald[18] wrote, prophetically: "*whisky now belongs not to the Scots but to the world at large*".

MacDonald's point was a philosophical one and he would have been surprised to learn of whisky being made in Corsica, or Nepal, or Brazil. But these are not whiskies in the Scottish style, or even the Corsican, Nepalese or Brazilian style, for there is no such thing. The new world whiskies have not arisen over hundreds of years of custom and practice but are transplants, for the most part the vision of a pioneer, inspired as often as not by a visit to an established distilling nation (generally, but not always, Scotland) to create their own personal interpretation of long-standing traditions.

However, the goal is rarely to ape established practice, for what would be the point? Local variations soon emerge, by accident or, more often, by design or in response to local conditions. Some distillers, such as Kavalan in Taiwan, are well-funded, can employ the best expertise and install the finest equipment, yet evolve their practices to meet local conditions (in this case, locally available yeasts and a maturation regime designed to deal with Taiwan's subtropical climate). Others, such as Mackmyra in Sweden, deliberately introduce juniper wood into the malting process and use Swedish oak casks to identify their products as Swedish whisky.

Between Zurich and Basle at the Whisky Castle, distiller Rudi Käser collects newly-fallen virgin snow, which he melts for the water to produce his unique Swiss snow whisky. His is a tiny operation, set up to diversify the family business in farming and the distilling of fruit spirits yet already carving a distinct identity.

In many cases, the distillery is strongly identified with the personality of its founder, or Master Distiller, and in extreme cases carries the same name; *e.g.* the Lark Distillery in Tasmania is named after Bill Lark, the founder. Another example is the Balcones Distillery in Waco, Texas where founder and distiller, Chip Tate, has cultivated a deliberately iconoclastic approach, beloved in the blogosphere, which blends personality, brand character and corporate identity.

As yet, world whiskies have no meaningful history, certainly nothing stretching back to medieval manuscripts or pioneering Scots and Irish settlers with memories of their home fires. In all probability many will fail – but this history is being written as we watch and taste.

1.6 LITERATURE

The commentary below relates solely to technical works, not titles aimed at the consumer published in English.

A number of early works were written and originally published in Latin and therefore fall outside the scope of this necessarily cursory review. Amongst the most important of these works is the *Liber de Arte Distillandi* of Hieronymus Brunschwig (1500, translated to English in 1527) and Philippe Ulstadt's *Coelum Philosophorum seu de Secretis Naturae Liber* (1525). Both are lavishly illustrated. There were a number of editions, reprinting and translations into other languages. In 1559, Peter Morwyn, a Fellow of Magdalen College, Oxford published an English translation of Konrad Gesner's

Treasure of Euonymus, which meticulously illustrates *"the formes of sondry apt fornaces, and vessels, required in this art".*

Early literature on distilling betrays the science's roots in medicine and alchemy and is sometimes mystical and allegorical in tone. Later works introduce place distilling in the context of the domestic or agricultural economy, demonstrating the small scale nature of much distilling in historic periods, before the literature becomes more scientific and specifically related to beverage alcohol.

Published in 1562, William Bulleins' *Bulwarke of Defence against all Sicknesse, Soarenesse, and Woundes that doe dayly assaulte mankind* is notable for being one of the first English herbals with a discussion of a number of mineral and chemical substances and methods of distillation, illustrative of the medicinal uses of *aqua vitae.*

Similarly, around 1634, Thomas Johnson translated from French View Larger Image*The Workes of that Famous Chirurgion Ambrose Parey,* an important medical text, the last section of which presents simple medicines, distillations and even instruction on embalming a corpse.

Around 1652, John French, a physician, published his *Art of Distillation,* essentially a medical reference and *The London Distiller,* describing *"all sorts of spirits and strong-waters: to which is added their vertues".*

Typical of the genre of works on domestic economy, *The True Preserver and Restorer of Health,* by G. Hartman, Chymist (1682) combined *"remedies for all distempers incident to man, woman and children. Selected from and experienced by the most famous physicians and chyrurgions of Europe together with excellent directions for cookery as also for preserving, and conserving"* and a wood cut illustrating the *"Engine for Distilling".*

Possibly the first work in English to deal with distillation as its principal topic was William Y'Worth's *The Whole Art of Distillation Practically Stated* (1692). Y'Worth was plainly unimpressed by French's work, considering his instructions *"defective, both in the exact Modus of working, the ordering of the wash and backs for a quick fermentation...as also in the great business of rectification".*

Nothing if not assured, Y'Worth's text is prefaced by an illustration of a distillery boldly captioned (Figure 1.7):

The Art of Distillation here behold
More perfect than before taught by tenfold.

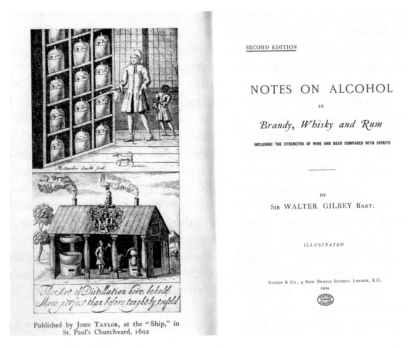

SECOND EDITION

NOTES ON ALCOHOL

IN

Brandy, Whisky and Rum

INCLUDING THE STRENGTHS OF WINE AND BEER COMPARED WITH SPIRITS

BY

Sir WALTER GILBEY Bart.

ILLUSTRATED

Vinton & Co., 9 New Bridge Street, London, E.C.
1904

Published by John Taylor, at the "Ship," in
St. Paul's Churchyard. 1602

Figure 1.7   The frontispiece of Y'Worth's *Whole Art of Distillation* (1692) as reproduced in Sir Walter Gilbey's *Notes on Alcohol* (1904). Note the verse under the illustration of the stillhouse, which reads "*the art of distillation here behold, more perfect than before taught by tenfold*".

Again, many of the recipes are for medicinal use but he also discusses cordial waters and "*the practical way of exalting malt spirits and how to endue them with flavours, measurably like those of wine*".

This was followed in 1718 by *The Practical Distiller* by an anonymous author, possibly George Smith of Kendal. Certainly, his *A Compleat Body of Distilling* (1725) was similarly printed by Lintot. This went on to become one of the two standard English works on distilling of the 18th century, enjoying five editions. It deals with the theory and practice of distilling liquors (including whisky) and cordials, intended partly for those larger houses that were equipped with a still-room.

A rival title, Ambrose Cooper's *The Complete Distiller* first appeared in 1757 but proved equally popular also running to five editions as late as 1826, by which time it had been re-titled

The Complete Domestic Distiller and was subtitled as being for the use both of *"Distillers and Private Families"*. Assuming Cooper to be 21 years of age at first publication, he would have been 90 at the final appearance of his work, which seems improbable; if so, it is a tribute both to his longevity and his ardent spirits.

Robert Shannon's *A Practical Treatise on Brewing, Distilling and Rectification* appeared in 1805 and may be considered the first modern book in English treating the subject. A large format volume, complete with illustrations, Shannon is credited with taking a recognisably systematic approach to the subject, aiming, as he says, to *"shew the distiller how he may proceed on rational principles"* (Figure 1.8). Detailed and practical directions are provided for the guidance of the distiller or brewer, together with tables showing, for example, what yield may be expected from a given quantity of grain.

Curiously for such a comprehensive work there appears only ever to have been one edition, despite the demand demonstrated by the repeated reprinting of the earlier titles by both Smith and Cooper.

A curious record of illicit distillation was preserved by Robert Armour, a plumber and still maker in Campbeltown, known to be

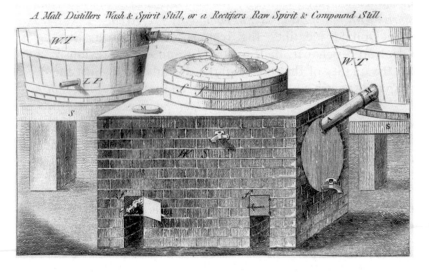

Figure 1.8 A malt distiller's wash and spirit still, from Shannon (1805).

working 1811–1817. Armour's notes of *Old Smuggling Stills* is a rare record of the systematic manufacture of stills for illicit distilling which, judging by the scale of his business, was widespread in Kintyre. Later, the firm supplied still safes to Springbank and Bowmore. For understandable reasons, it would seem to have been unusual for any record of any part of the trade in illicit distilling to have been recorded, let alone kept for any length of time. However, Armour's notes survive as a manuscript in a private collection, although, so far as we can establish, they have never been reproduced; however, in 1970, they were reviewed by Dr I. A. Glen*** in the journal, *Scottish Studies.*[19]

So far as we can determine, the earliest work written and printed in Scotland on distilling is by John McDonald, whose *The Maltster, Distiller, and Spirit Dealer's Companion* was published in Elgin in 1828. We have not seen this title cited in any of the major works on Scotch whisky, nor can we find it in any bibliography on the subject, from which we can deduce that it is rare, possibly exceptionally so. McDonald is described as 'of the Excise' and, in the preface, he acknowledges the encouragement from his colleague John Anderson, a Collector. The book is meticulous and detailed, with much on the economics of distillery operations and the sale of whisky and a thoroughly modern concern with quality.

On malting McDonald writes *"the malt from which the favourite mountain dew has been made, was generally dried with peat, and the spirits with that flavour is still commonly preferred"*.

McDonald identifies the regions of Fairntosh, Strathglass, Glenlivat, Lochaber, Badenoch and Rannoch as previously famed for the quality of their illicit spirits. He advises against the use of soap in the wash still, suggesting instead that *"suet, butter, or hog's lard"* would serve to avoid soap's *"pernicious taste"*. All three are commodities that presumably would have been readily available in a rural economy. He is also at pains to stress that *"before any thing is done in the distilleries, cleanliness above every thing, must, at all times, be observed"*.

As the earliest work on whisky so far identified, written and published in Scotland, McDonald's book is of great historical importance and deserves to be considerably better known and studied.

***Dr, then Miss Glen also notably contributed a scholarly introduction to the first facsimile reprint of Alfred Barnard's *The Whisky Distilleries of the United Kingdom* (David & Charles, 1969).

Although the French *Maison rustique, or The Countrie Farme* had appeared in a number of English editions as early as 1606 (including appetising instructions for the distillation of the blood of a male goat and even *"a man's dung"* – we have no idea why), a greater number of titles on agriculture appear in the early 18th century, in which distilling features and is treated as a natural extension of good farm management and part of the economy of a well-run farm. Amongst such titles we may include Nathan Bailey's *Dictionary of Husbandry, Gardening, Trade, Commerce & Country Affairs* (pre-1717), Professor Richard Bradley's *The Country Housewife and Lady's Director in the Management of a House, and the Delights and Profits of a Farm* (c. 1727) and part one of Bettesworth and Hitch's *The Complete Family Piece and Country Gentleman, and Farmer's Best Guide,* which, along with instructions on *"distilling and fermenting of all compound simple waters and spirits"* includes Dr Mead's *"cure of the bite of a mad dog"*[†††] (3rd edition, 1741).

Domestic distilling also formed a branch of cookery. A particularly early example from 1671 is N. Brooke's *A Queene's Delight* with a description of *"distilling the most excellent waters".*

Sarah Harrison's *The House-Keeper's Pocket-Book, and Compleat Family Cook* first appeared in 1733 but shortly ran to several expanded editions. Along with several hundred recipes and medical advice, it described *"distilling strong waters, &c. after the most approv'd method".*

Probably written under a pseudonym, Arabella Atkyns' The Family Magazine (1741) purported to offer "several hundred receipts [recipes] in cookery, pastry, pickling, confectionary, distilling, brewing, cosmeticks, &c. together with the art of making English wines" and went on to offer a "compendious body of physick; succinctly treating of all the diseases and accidents incident to men, women, and children". The similarity to Hartman (1682) is unlikely to be coincidental.

Elizabeth Raffald, the Housekeeper to Lady Warburton, published her *The Experienced English Housekeeper* in 1794, including, amongst the claimed 900 original recipes, two for distilling, including instructions on refining malt liquors.

[†††]One can never assert with certainty when such guidance may prove of assistance. Perhaps all future titles on distilling should carry similar intelligence.

By 1808, Duncan Macdonald, the Head Cook at the Bedford Tavern and Hotel, Covent Garden, could offer *The New London Family Cook; or, Town and Country Housekeeper's Guide,* which was described as: *"comprehending directions for marketing... practical instructions for preparing soups, broths, gravies, sauces, and made dishes... with the respective branches of pastry and confectionary, the art of potting, pickling, preserving, &c., cookery for the sick, and for the poor; directions for carving... Also a collection of valuable family recipes, in dyeing, perfumery, &c. Instructions for brewing, making of British wines, distilling, managing the dairy, and gardening. And an appendix, containing general directions for servants relative to the cleaning of household furniture, floor-cloths, stoves, marble chimney-pieces, &c., forming in the whole a most complete family instructor"*.

Volume I of the *Cabinet Cyclopædia* of 1830, entitled *Domestic Economy* by Michael Donovan, Professor of Chemistry in the Company of Apothecaries of Ireland, expounds upon distilling, giving recipes and instructions in copious detail, along with a brief history of intoxicating liquors and spine-tingling tales of spontaneous combustion in habitual drunkards.

As many of the cookery titles were reprinted several times without significant revisions, references to domestic distilling carry on well into the Victorian period.

However, a curious late survival of this genre is *The Still Room* by Mrs Charles Roundell and Harry Roberts, published in London by The Bodley Head in 1903 as a stand-alone volume but described as volume IV of *The Country Handbook* series. As well as a number of illustrations of more or less fantastical medieval stills, a practical looking 'portable copper still' is illustrated and its operation briefly described. Distillation is described thus: *"there is no occupation that comes nearer to the work of gods than this occupation of distilling"* and in terms reminiscent of alchemy the author continues: *"the distiller can but smile at the impotence of those who are unable to conceive the possibility of a post-physical human existence, for, day by day, as he stands before his stills he sees the miracle performed whereby the spiritual, the essential, is separated and continues to exist apart from the material body in which it previously dwelt"*.

Mrs Roundell gives a number of recipes, including various cordials: 'Benedictine', 'Green Chartreuse', 'Kummel' and absinthe, all involving the redistillation of rectified spirits. Recipes for usquebaugh and the delightfully named 'Sighs of Love' are essentially infusions, not involving any further distillation. The significance of this title seems to us to lie in the suggestion that distilling was fairly openly carried on in respectable houses as a branch of cookery as late as 1903 and thought unremarkable.

In Edinburgh, The Society of Improvers in the Knowledge of Agriculture in Scotland, founded in 1723 and presided over by Thomas Hope of Rankeillor, who had studied agriculture both in England and on the European continent, promoted the adoption of improved farming methods. Their *Transactions* (published in 1743) include a discussion of distilling as an adjunct to the farm. Similarly, subscription volumes, such as the *Annals of Agriculture* and the early 19th century *Farmer's Magazine,* aimed at the progressive landlord or improving tenant, from time to time carried information on distilling.

Distilling had now taken root in the American colonies and a number of textbooks appeared. First amongst these was *The American Distiller* (Philadelphia, 1804) by Michael Krafft who evidently worked as a consultant to a number of operations, stating in his advertisement (*i.e.* the preface) that there were "*not less than two hundred and seventeen distilleries working on my plan*" and that he was willing to "*attend the erecting of stills or distilleries...on terms that will sufficiently meet the interests of the proprietors.*"

Note the date of Krafft's work. In her powerfully suggestive, albeit geographically constrained, study of early distilling in colonial Chesapeake,[12] Sarah Hand Meacham convincingly demonstrates that "*making alcoholic beverages was something that women did as part of their share of the household labor*"; that "*making alcohol used to be women's work*" and that it was primarily the arrival of technology that permitted men to claim "*that making alcohol was a science that belonged within men's domains*". As Meacham goes on to show, the role of women in early distilling practice has largely been ignored and that "*by the early twentieth century, most American women and men had forgotten that making alcohol was once part of women's cookery*". No doubt, much the same could be said of the UK.

Notwithstanding Krafft's claim of more than two hundred satisfied clients, only five years later, Samuel McHarry (the title page reads M'Harry, presumably because the printer lacked the correct type) published his *The Practical Distiller* (Harrisburgh, PA; 1809) claiming in the preface that "*I thought there must be books containing instructions, but to my surprise, after a diligent search of all the book-stores and catalogues in Pennsylvania, I found there was no American work extant, treating on this science...*". Perhaps his search lacked some diligence or possibly he and Krafft were commercial rivals, but this sentence has led a number of commentators to conclude that McHarry's was the first work published in America, whereas this honour rightly belongs to Michael Krafft.

The third early American work is *Hall's Distiller* (Philadelphia, 1813) by Harrison Hall, which concludes with a seven page "*list of patents granted by the United States, for improvements in distillation, on stills, and refining of liquors*" covering the period 1791–1812.

Later American works include *The Art of Making Whiskey* (Lexington, 1819 by Anthony Boucherie); *The Complete Practical Distiller* (1853) by Marcus La Fayette Byrn; Lewis Feuchtwanger's *Fermented Liquors* (1858, an early example of self-publishing); *The Brewer, Distiller and Wine Manufacturer* (1883) by John Gardner; Joseph Fleischman, of the famous distilling family, with *The Art of Blending* (1885); *The Practical Distiller* (1889) by Leonard Monzart and, following the repeal of prohibition, *The Manufacture of Whiskey, Brandy and Cordials* (1934) by Irving Hirsch.

So far as the UK is concerned, there appears to be a gap of nearly 50 years in the technical literature after MacDonald's rather rare work of 1828. The next relevant title that we can identify – and a highly specialised one at that – is the *Tables of Spirit Proofs*, published in Edinburgh, 1877 and compiled by Duncan McGlashan of the city's Caledonian Distillery,[‡‡‡] a large grain whisky operation.

The *Tables* are, as the name suggests, exactly that – several hundred pages of figures allowing the quantity of spirits at a given proof to be easily determined.

In 1878, the four principal Irish distillers (John Jameson and Son; Wm. Jameson & Co.; John Power and Son and George Roe & Co.) issued a powerful and vehement denunciation of the trade in 'silent spirit' or 'sham whisky'. *Truths About Whisky* (note the spelling) was part of a sophisticated lobbying and P. R. campaign

---

[‡‡‡]At one time, the 'Caley' as it was known, was one of the largest distilleries in Scotland. Little remains today, save for a chimney and some buildings. Most of the site has been redeveloped for housing.

to defend the reputation of Dublin pot still whisky from the impact of blending and from 'passing-off' and counterfeit Irish whiskies. The campaign ultimately failed, but the book is a wonderful polemic giving a valuable insight into some of the darker and more nefarious practices of the late-Victorian whisky trade. Without it we might have forgotten all about 'grogging' and the charms of 'Hamburg sherry', 'prune wine' and 'cocked hat spirit'.

In 1887, Alfred Barnard's magnum opus *The Whisky Distilleries of the United Kingdom* appeared, a collection of his reports for *Harper's Weekly Gazette*, and this has been reprinted on a number of occasions since as the main reference source for Victorian distilling in its heyday.§§§

It is not, however, a working guide to operating a distillery. That arrived in 1896 with the publication of J. A. Nettleton's first major work, *The Manufacture of Spirit in the United Kingdom*, which he subsequently revised and updated just prior to World War I.

In 1903, Sir Walter Gilbey, then owner of the Glen Spey, Strathmill and Knockando distilleries, published a slim pamphlet, titled *Notes on Alcohol*, which strongly advocated the merits of direct firing and pure pot still whisky. *"When a plain, flavourless spirit, the produce of the patent still, is blended with a pot still whisky"*, he wrote *"it is but a very poor compromise"*. However, as the history of the firm records,[20] by 1905 *"it was gradually appreciated that malt whisky was too heavy for a southern climate"* and grain whisky was admitted.

Gilbey was writing against the background of an increasingly heated debate on the nature of whisky. Easily the most significant publication of the 12th century was the 1909 report of the Royal Commission on *Whiskey and other Potable Spirits* (again, note the spelling of whisky). This followed a long period of public debate,¶¶¶ culminating in the famous 'What is Whisky?' question arising from the 1905 test case prosecution in Islington Magistrates Court of two unfortunate publicans accused of supplying both Irish and Scotch whiskies not *"as demanded by the purchasers"*. Found guilty, they immediately appealed and, after lengthy machinations

---

§§§Some years later, Barnard wrote a number of commissioned pamphlets, probably six in number, for different companies, which go into greater detail than the fairly terse entries in his book. These, however, are now very rare.

¶¶¶See, for example, the pamphlet *The Practice of Substitution in the Spirit Trade* reproduced in *The Lancet* in February 1903. There was also extensive correspondence in the press.

in which William Ross of the DCL played a prominent role, a Royal Commission was announced in February 1908.

Its determinations, by allowing grain spirit to be used in blending and accepting the result as whisky, laid the foundations of the modern industry. Most in the Scotch whisky trade were satisfied with the result, though not all and it certainly did not appeal to the traditionalists in Irish distilling. Amongst other prominent critics was J. A. Nettleton, who wrote: "*the opinions of medical men cannot be cited as expert evidence in regard to whisky*" – a pointed reference to the make-up of the Royal Commission, which was heavily weighted in favour of the medical profession. But, regardless of criticism, the conclusions of the Commission held and went on to determine whisky's future course.

Nettleton, a former Inland Revenue analyst, had previously published on *The Flavour of Whisky* (1894 – this may have been an article in a trade journal; we have been unable to definitively establish any detail on this) but the work for which he is best remembered is the successor to his earlier work *The Manufacture of Whisky and Plain Spirit* (Aberdeen, 1913). This remained the standard work for a number of years afterwards and, in some ways, has never been entirely replaced or superseded – to our knowledge, it was consulted by a well-known Irish distiller as recently as 2011 as they endeavoured to re-start and re-use a vintage column still.

In 1937, William Birnie, an accountant and Charles Mackinlay's partner in Glen Mhor and Glen Albyn distilleries circulated a slim typescript entitled *Notes on the Distillation of Highland Malt Whisky*, which was later printed as a small booklet by the Glasgow firm of blenders, Brodie Hepburn (1964).

Apart from this latter edition, however, it is beyond the scope of this review to consider work published after the World War II. It may be regarded as a point of interest that in recent years there have been a number of works published to guide the small-scale distiller, presumably in response to the boom in craft distilleries.

Finally, it should be noted that there are many valuable and useful books on such as cooperage, malting and gauging. These have not been considered here but their omission demonstrates a need for a thoroughly comprehensive bibliography of titles related to whisky, distilling and related subjects.

REFERENCES

1. P. Levi, *The Periodic Table*, Michael Joseph, London, 1985.
2. H. Brunschwig, *The Vertuose Boke of Distyllacyon*, 1st English ed., Laurens Andrewe, London, 1527.
3. F. Greenaway, *et al.*, in *Medieval Distilling, J. Medieval Archaeology*, vol. 16, London, 1973.
4. F. Moryson, *An Itinerary Written by Fynes Moryson Gent.*, John Beale, London, 1617.
5. E. B. McGuire, *Irish Whiskey*, Gill & Macmillan, Dublin, 1973.
6. A. Barnard, *The Whisky Distilleries of the United Kingdom*, Harper's Weekly Gazette, *London*, 1887.
7. Anon., *Truths about Whisky*, Sutton Sharpe & Co, London, 1878.
8. L. Brown, *The Story of Canadian Whisky*, Fitzhenry & Whiteside, Markham, 1994.
9. D. de Kergommeaux, *Canadian Whisky*, McLelland & Stewart, Toronto, 2012.
10. G. Regan and M. Haidin Regan, *The Book of Bourbon*, Chapters Publishing, *Shelburne*, 1995.
11. S. Morewood, *Inebriating Liquors*, William Curry, Jun & Company and others, Dublin, 1838.
12. S. Hand Meacham, *Every Home a Distillery – Alcohol, Gender and Technology in the Colonial Chesapeake*, Johns Hopkins University Press, Baltimore, 2009.
13. T. R. Dewar, *A Ramble Round the Globe*, Chatto & Windus, London, 1894.
14. G. Blake, *Correspondence with Frank Morley of Faber & Faber*, unpublished, 1930.
15. O. Checkland, *Japanese Whisky Scottish Blend*, Scottish Cultural Press, Edinburgh, 1998.
16. Ocean's early years (Part I) – Daikoku Budoshu and the birth of the brand, 2012. Available at http://nonjatta.blogspot.jp/2012/03/oceans-early-years-part-i-daikoku.html (accessed 24th July 2013).
17. History of Japanese Whisky Exhibit, 2013. Available at http://www.suntory.com/factory/whisky/exhibition/index.html (accessed 24th July 2013).

18. A. MacDonald, *Whisky*, The Porpoise Press, Edinburgh, 1930.
19. I. A. Glen, *A Maker of Illicit Stills*, in *Scottish Studies*, vol. 14, part 1, School of Scottish Studies, University of Edinburgh, 1970.
20. A. Waugh, *Merchants of Wine*, Cassell and Company, London, 1957.

CHAPTER 2

Scotland

The standard history of distilling in Scotland begins with Friar John Cor of Lindores Abbey in Fife and the reference to him in the Exchequer Rolls of 1494, in which he is recorded as receiving "*eight bolls of malt*". The Exchequer, one of the earliest government departments, developed out of the King's Chamber, the branch of the royal household which oversaw the royal finances at a time when Scotland was an independent kingdom. The chief financial officer was the King's Great Chamberlain and the Exchequer was not a permanent body, meeting only to audit the accounts of the sheriffs and other collectors of royal revenues.

James I reformed the Exchequer in the 1420s. Its functions were divided between the Comptroller (or Receiver General) and the Treasurer. The Comptroller handled the revenue from Crown lands, burghs and customs, which was spent on the royal household. The Treasurer received the feudal services and casualties (occasional payments to a superior of lands), the proceeds of taxation and the lucrative profits of justice. From these revenues he met the King's personal expenses, including military and naval expenses, liveries, stables, repair of palaces and alms.

The significance of whisky as a source of revenue has not escaped the attention of subsequent governments and has attracted much debate over the years. It is entirely fitting then that the first

The Science and Commerce of Whisky
By Ian Buxton and Paul S. Hughes
© Buxton & Hughes, 2014
Published by the Royal Society of Chemistry, www.rsc.org

recorded mention of whisky in Scotland can be found in a record of taxes.

As noted, Friar John Cor is recorded as receiving "*eight bolls of malt*". A 'boll' was an old Scottish measure of weight, equivalent to 6 bushels of malt or approximately 140 lbs (63.5 kg). Thus eight bolls is around 0.4 of a tonne of malt[†] and this substantial quantity of malt, combined with the laconic tone of the ledger entry, strongly implies that this was not the first time that *aqua vitae* had been produced. As a present day distiller could theoretically produce as much as 1500 bottles of whisky[‡] from this quantity of raw material we may be confident, therefore, that the origins of Scotch whisky lie somewhere earlier in history and perhaps an earlier record may yet emerge.

Claims, in fact, that are sometimes made for Michael Scott., known as the so-called 'Wizard of Balwearie', who trained at Oxford, Paris and Toledo universities in the early 13th century. Manuscripts attributed to him refer to 'Aqua Ardens', the earliest name for distilled alcohol. Returning to Scotland around 1230 he may have brought with him distilling secrets and he would certainly have been familiar with alchemy. Could he be the father of Scotch whisky?

Whatever the romantic appeal of Scott's case, records of a distilled alcoholic spirit dating from ca. 1150 come from the famous medical school at Salerno in southern Italy, and from the University of Toledo in Spain. There is general agreement that the art of distilling came to Scotland with Irish monks. Lindores was a Tironensian abbey, founded around 1191 and eventually sacked in the late 16th century. Despite its apparent significance to Scotch whisky, the industry has done little or nothing to promote or restore it. Although grand-sounding plans surface from time to time,[§] these have until now proved abortive.

Three years after the reference to Friar Cor, King James authorised a payment "*to the barbour brocht aqua vitae to the King in Dundee, be the Kingis command*" and, in 1505, the Guild of Surgeon Barbers in Edinburgh was granted a monopoly over the

---

[†]Alfred Barnard suggests a boll was equal to 240 lbs weight, thus 870 kg in total.
[‡]We know little or nothing of the efficiency of medieval distilling apparatus, the yields obtained from a given quantity of malt or the skill of the distillers or the strength to which the spirit was distilled. Any estimates of output are, therefore, entirely theoretical and a matter of conjecture.
[§]See, for example, http://ww2.duncantaylor.com/products/lindores.htm.

manufacture of *aqua vitae*.[¶] There are in fact 15 further references to *aqua vitae* in the Exchequer Rolls from 1494 to 1512 with James, who was an enthusiastic supporter of the Surgeons. He wasn't drinking all this however. The records clearly show that most was destined for his experiments in chemistry and medicine and, possibly, for improving the performance of gunpowder. Monastic distilling may also have been for embalming fluid, disinfectant or for cosmetic use. We simply don't know.

By the 1550s infringements of the monopoly were occurring with some regularity and, in 1555, the Scottish Parliament restricted distilling to burghs in the west due to a failure of the harvest. A further act of 1579 restricted distilling to Earls, Lords, Barons and Gentlemen, again in response to a shortage of grain for food, implying that distilling had become widespread. Such restrictions were to become relatively commonplace over the next 200 years.

In 1644, an excise duty of 2 shillings and 8 dimes per Scots pint was fixed by the Scottish Parliament on spirits offered for sale. However, much distilling was on an entirely domestic scale and, although the produce may have been bartered or used to pay agricultural rents, it was not undertaken as a commercial enterprise. Indeed, distilling seems to have been so commonplace a part of everyday life that few commentators thought it worth remarking upon. The existence of legislation to tax whisky proves the existence of larger-scale, commercial operations but it is not until 1689 that we can identify even one of them.

That is the Ferintosh distillery (also known as Ryefield) of Duncan Forbes, whose enthusiastic adherence to William of Orange so enraged the Jacobite supporters of King James VII of England and II of Scotland that they burnt it to the ground. It must have been a substantial operation as his son claimed the staggering sum of £54,000 in compensation. He was awarded the privilege of distilling whisky free of duty from grain grown on his own land – and so profitable did this prove that he and his descendants were able to expand the estate and eventually build three further distilleries.

In 1784, the concession was eventually revoked and the family received a further £21,580 in compensation. The distilleries closed

---

[¶]The monopoly continued until 1757; however, by then, it was honoured very much more in the breach rather than the observance. It appears to have effectively lapsed around 1612, the last appearance of *aqua vitae* in the records of the College of Surgeons.

the following year and Robert Burns famously lamented their disappearance in his poem *Scotch Drink:*

> *Thee, Ferintosh! Oh, sadly lost!*
> *Scotland lament frae coast to coast!*
> *Now colic grips, and barkin' hoast*
> *May kill us a';*
> *For loyal Forbes's charter'd boast*
> *Is ta'en awa'!*

The Ferintosh district still appears on the map accompanying the 1799 Parliamentary Report on the distilleries in Scotland and, in 1805, Robert Forsyth, writing in *The Beauties of Scotland*, noted that Ferintosh whisky *"was much relished in Scotland"*.

But that is to jump ahead. There are sporadic references to *aqua vitae* in travellers' accounts of Scotland – all too frequently interspersed with incredulous accounts of the levels of consumption and patterns of drinking – that suggest distilling was widespread and that the product was consumed with enthusiasm.

By 1707, Scotland was effectively bankrupt following the failure of the Darien Scheme. The Act of Union offered a way out – wider access to markets and the debt written off. Although widely criticised, then and since, it is hard to see what realistic alternative was open to the Scottish Parliament, which duly acquiesced in its own extinction. However, despite the treaty uniting the parliaments fixing identical duties on excisable liquor in both countries, Westminster imposed a malt tax in 1713, which was increased in 1725. This was highly unpopular but paradoxically had the effect of increasing the production and consumption of spirits in Scotland, which was further boosted by the 1736 Gin Act. The malt tax did, however, contribute to a growth in Jacobite sympathies and there were Risings (or Rebellions, depending on your point of view) in 1715 and 1719, precursors of Bonnie Prince Charlie's disastrous 1745 adventure, and about as effective.

Smuggling and illicit distillation greatly increased after 1707. Sir Walter Scott later wrote: *"smuggling was almost universal in Scotland during the reigns of George I and II,[||] for the people, unaccustomed to imposts and regarding them as an unjust aggression*

[||] George I, 1714–1727; George II, 1727–1760.

upon their ancient liberties, made no scruple to elude them wherever it was possible to do so". MacLean[1] notes the romantic image of the smuggler as being widespread due to a general acceptance or 'amused tolerance' amongst almost all classes of society of the smuggler as a 'local hero'.

Notwithstanding this benign view, smuggling was, on occasion, accompanied by considerable violence and more than one excise officer, or gauger, was killed in the execution of their duties: a parliamentary enquiry recorded six deaths between 1723 and 1736 and some 250 officers *"beaten, abused and wounded"*. The draconian penalties for smuggling or illicit distillation, and the substantial profits to be made were a powerful incentive to avoid capture and many organised gangs of smugglers were heavily armed and ready to defend themselves at all costs.

The Robin Hood image of the smuggler owes much to an over-romanticised view of their activities, although grinding poverty and the lack of viable alternative employment played a part in turning honest men and women into law-breakers. This situation was to reach a memorable height of absurdity during the state visit of King George IV to Scotland when the monarch was to drink – with conspicuous enjoyment – quantities of illicitly distilled and smuggled Glenlivet supplied by Elizabeth Grant on behalf of her father.

The consumption of much whisky, along with the flamboyant wearing of extravagant tartan costume and the deliberately meticulous planning of the formal events on the King's tour were elaborately orchestrated by the writer, Sir Walter Scott. Much of the imagery and iconography that today is associated with Scotland was Scott's deliberate creation and has, in many ways, proved a greater and more lasting legacy than his historical novels.

As Elizabeth Grant recalled in her memoirs,[2] *"Lord Conyngham, the Chamberlain, was looking everywhere for pure Glenlivet whisky; the King drank nothing else. It was not to be had out of the Highlands. My father sent word to me – I was the cellarer – to empty my pet bin, where was whisky long in wood, long in uncorked bottles, mild as milk, and the true contraband goût in it"*.

Elizabeth Grant is being a trifle disingenuous: other commentators are agreed that the King was more than partial to cherry brandy. He does not appear to have been a particularly discriminating drinker but, while George's enthusiastic consumption of Glenlivet has become embedded in whisky's mythology, memories of his other 'wee refreshments' appear to have dimmed. Her praise for whisky 'long in wood' is clear enough but the reference to 'long in uncorked bottles' is puzzling – possibly this was to allow harsher or more volatile alcohols to evaporate.

She also goes on to describe in somewhat sanctimonious detail, the Highlanders' relationship with whisky:

> *"At every house it was offered, at every house it must be tasted or offence would be given, so we were taught to believe. I am sure now had we steadily refused compliance with so incorrect a custom it would have been far better for ourselves, and might all the sooner have put a stop to so pernicious a habit among the people. Whisky-drinking was and is the bane of that country; from early morning to late at night it went on. Decent gentlewomen began the day with a dram. In our house the bottle of whisky, with its accompaniment of a silver salver full of small glasses, was placed on the side-table with cold meat every morning. In the pantry a bottle of whisky was the allowance per day, with bread and cheese in any desired quantity, for such messengers or visitors whose errands sent them in that direction. The very poorest cottages could offer whisky; all the men engaged in the wood manufacture drank it in goblets three times a day, yet except at merry-making we never saw anyone tipsy".*

During the 18th century there were considerable improvements in agricultural technology and land management, leading to increased crop yields, greater prosperity and an excess of grain that could be distilled – turning a bulky, perishable, low-value crop into something easily stored and transported, not attractive to vermin and of higher value. Accordingly, the earliest known commercial distilleries start to appear, often promoted by progressive landlords and examples include Kennetpans (c. 1720), Cambusbarron (1741), Dolls (1746), Portree (1752), and so on. By 1751, Cambusbarron was exporting to England and overseas, their duty payments accounting for 1% of the total duty on British spirits in Scotland.

Around this time 'whisky' starts to appear. Moss and Hume[3] cite a reference in correspondence from 1736, in which Baillie Steuart of Inverness implores his brother-in-law to "*forbear drinking that poisonous drink. I mean drams of brandie and whisky...*". However, also in 1736, Captain Edward Burt of General Wade's staff relates that "*some of the Highland gentlemen are immoderate drinkers of Usky...*". The terms seem to have been inter-changeable but, within a few years, meanings had stabilised: 'usquebaugh' appears to denote compounded liquor (refer to Johnson's Dictionary of 1755, although he then confuses matters by equating it with whisky) and '*aqua vitae*' and '*uisge beatha*' indicate malt spirit.

By 1759, 'one of the members' of the Ratho Club (said to be James Reid of Long Hermistoun) published "*An Apology for Whisky*",[4] apparently confident that it would be understood as referring to *aqua vitae* or malt spirits. He deplores "*the prostitute use of tea and the profuse drinking of wine among people of condition*" and concludes "*upon the whole, the distilling of whisky is a national benefit, it has not corrupted the country with drunkenness, but reformed it, our trade and manufacture never flourished more than during its subsistence, and by its assistance; and the condition of the poor is happier, as they are better employed. By means of it, many of our gentlemen live in towns and ride in chariots; by means of it our farmers eat bread, and pay their rents, their merchants and their servants; and according to the import of its name, it is really, in our present circumstances, aqua vitae.*"

Perhaps he was being ironic. Due to crop shortages, the government had banned distilling from 1757 to 1760 and many of the fledgling licensed distilleries were forced to close. Domestic distilling remained legal, so long as the product was not for sale; illicit distillation and smuggling boomed and Duncan Forbes in his Ferintosh tax-free zone continued untroubled by the excise. Naturally, he expanded his operations, which presumably began to fuel the resentment against him felt by his tax-paying competitors.

Many commentators refer to Hugo Arnot's 1779 *History of Edinburgh,* in which he asserts that the city held "*no fewer than 400 private stills which pay no duty to government*" contrasting this with licensed stills "*which indeed are only eight in number*". However, he goes on frankly to admit of the figure of 400 that "*this estimate, however, is only conjecture*" and, as he was fervent in his desire to stamp out whisky drinking, we are not sure that he can be admitted as a reliable or balanced witness.

Arnot, who comes across as something of a killjoy, describes whisky as "*pernicious to health and to morals*" and suggests that "*every means ought to be used for lessening the manufacture of whisky*". Not content with that, he suggests that "*private families ought to be prohibited from distilling spirituous liquors for their own consumption, as being a cloak to perjury, defrauding of government, and drunkenness"*. He was no more enthusiastic about the *"lower class of inhabitants"* having *"betaken themselves to tea"*. How scandalous.

However, Arnot was writing at a time when production had boomed, especially in the Lowlands of Scotland. At least 23 new distilleries, some of them very large, were built and soon prospered on the back of a profitable export trade to England, where their spirit was rectified and made into gin. No doubt he was gratified by tax increases and restrictions on distilling in 1779, 1781 (when private distilling was outlawed) and 1783. During the last two decades of the 18th century, we see a flurry of acts, culminating in a comprehensive Parliamentary Report of 1798/1799. Amongst the most important of these was the 1784 Wash Act, revised and amended in the following year. Its provisions included the withdrawal of the privileges at Ferintosh.

There had been great privations in 1783 and 1784 due to the failure of the harvest, and distilling was restricted. However, the so-called 'great and middle class distilleries' of the Lowlands, who to a large extent were anyway working with ingredients inferior to the markedly smaller Highland stills, imported inedible and poor quality grain from the Continent and continued distilling and shipping spirit to England. This trade was resented by the London distillers who maintained that they had been disadvantaged. Not all commentators were impressed. An anonymous commentator on Sir John Dalrymple's proposals for levying duty, for example, detected in their motives *"a supreme degree of selfishness, and a mean jealousy of the rivalship of the Scottish distillers"*.[5]

But others simply took direct action, so desperate was their condition. In June 1784, a crowd gathered at James Haig's Canonmills distillery (Edinburgh) intent upon looting the grain and vegetables they believed were stored there. The mob was dispelled only after one rioter was shot by the militia; two ringleaders were publicly whipped and sentenced to 14 years transportation.

The practical and immediate effect of the Wash Act and its amending act was to provide the great and middle class distilleries with an incentive to increase production, an opportunity that they seized immediately, introducing a number of technical innovations that enabled them to discharge their stills at the almost inconceivable rate of once every 2 minutes and 45 seconds.

Production immediately boomed, with significant quantities being sent to England, where it was made into cheap gin. Retribution from the London distillers, well organized and powerful, followed in short order.

The scale and impact of the great distilleries at Kilbagie and Kennetpans should not be under-estimated. These were giant concerns, even by today's standards and at the forefront of the Industrial Revolution in Scotland. Both distilleries had in excess of 850 acres of farmland at their disposal and Kilbagie alone generated enough animal fodder to fatten 7000 cattle and 2000 pigs. It had a staff of over 300 directly employed on site, not including the many other ancillary jobs connected to distilling.

Some of Scotland's earliest coal mines were located at Kennetpans, which also housed the first James Watt steam engine installed in the country. The canal linking Kilbagie and the port at Kennetpans was over a mile in length and remained in use until 1861, while the two distilleries were linked by Scotland's first railway line, and Kilbagie employed Scotland's first threshing machine from 1787.[6] It is hardly exaggerating to assert that the historical significance of Kennetpans and Kilbagie to the history of distilling, even if little appreciated, can hardly be over-stated. This is the crucible in which the modern Scotch whisky industry was formed. Sadly, though, much of the site remains just about intact, the buildings at Kilbagie are now rapidly deteriorating. A charitable trust has been formed with the aim of restoring and conserving what can be saved.

During 1786 some 881 969 gallons (just over 4.0 million litres) of whisky were sent duty paid to England to the chagrin of the London distilling trade. Their vigorous lobbying resulted in the Scotch Distillery Act, imposing the penal duty of 2 shillings per gallon on spirit exported to England.[**] This was further increased in 1788 and later that year the Lowland License Act required distillers working for the English market to give 12 months' notice of their intention to do so. The Lowland distilling industry collapsed; the Stein and Haig families were bankrupted (though they appear to have made prudent provision and were shortly to repurchase the businesses) and a huge quantity of poor quality, raw and immature spirit swamped Scottish markets.

[**]As a result of this one-sided legislation, Robert Burns wrote the passionate *The Author's Earnest Cry and Prayer to the Scotch Representatives in the House of Commons:"freedom an' whisky gang thegither!"*

A contemporary pamphlet[7] described unmalted corn spirits *"sent into the market, smoking hot from the still, in such a state, that the person must be possessed with the fortitude of Socrates, who can swallow the contents of a cup without having the muscles of his face distorted with convulsions"*.

Yet more lobbying and legislation followed. Smaller legal distilleries were driven out of business by the excess of supply of cheap Lowland spirit and considerable quantities of better quality smuggled Highland whisky (by accident, design or connivance the excise officers appear to have been remarkably ineffective). By 1793, to fund the war against France following the French Revolution of 1789, taxes were again increased and, with support from family connections and backed by their creditors, the Stein and Haig dynasty reappeared *"like bad pennies"*.[8]

Distilling (legal and licensed that is, for illicit production continued largely unabated) was suspended in 1795 due to two bad harvests and duty was increased, only for further impositions to come into force when the restrictions were lifted. An unknown poet hailed the resumption of distilling in a chapbook entitled *Cheap Whisky, a Familiar Epistle to Mr. Pitt on the Recommencement of Distilling in Scotland. To which is Added, The Gowd O' Gowrie: A Scots Song Never Before Published,*[9] prefaced by the comment that *"an unusual portion of joy has been thereby diffused among the lower classes in Scotland, who indulge the pleasing hope of again tasting their favourite beverage, the high price of which, had almost amounted to a total prohibition"*.

> *"Of a' the ills in this creation*
> *Drouth [thirst] and nae drink's the warst vexation"*

Curiously, the prohibitions on distilling at this time are also recorded in popular culture through the composition in 1799 by the renowned Perthshire fiddler, Neil Gow, of the mournful lament *A Farewell to Whisky*. Fortunately, two years later he was able to compose the altogether jauntier Strathspey, *Whisky Welcome Back Again*. Both remain a vibrant part of the folk repertoire.

The fast working of the shallow Lowland stills caused the wash to stick to the base of the still and burn − to the detriment of the quality of the spirit. Sometime during the 1790s, this problem was addressed by the invention of the rummager, a device which revolved round the still, scouring the sides and base preventing particles sticking and, in addition, releasing small amounts of copper, which had the unintended, though beneficial, effect of reducing the sulphur content of the spirit. The rummager remains in use to this day in direct-fired stills.

Frustrated by the state of the distilling industry and no doubt by now thoroughly confused by the lobby and counter-lobbying of the various interests and their wildly varying and contradictory claims and counter-claims, the government set up a Select Committee of the House of Commons in 1797 with a brief *"to enquire into the best mode of levying and collecting the duties upon the distillation of corn spirits in Scotland"*.

Under the chairmanship of the Rt Hon. Sylvester Douglas, whose energy one can only marvel at, evidence was taken, witnesses called and examined, excise records poured over and two substantial reports (June 1798 and July 1799) produced. They are a remarkable, detailed and lively record of the state of the industry at the end of the 18th century, wonderfully illustrated and with a wealth of information. The Committee heard from every figure of substance in the industry and received a mass of contradictory submissions and evidence from interested parties. It is worthy of detailed analysis in its own right.

Typical of the submissions to the Select Committee was that of the Rev. George Skene Keith, then minister at Keith Hall and Kinkell, Aberdeen. Keenly interested in agricultural questions and a stout defender of the Highland distiller, Keith commented sardonically on the complaints of the Lowland distillers. *"We are told that a certain Lowland distiller and his family were so 'completely ruined' by the Ferrintosh [sic] privilege, that his eldest son was thrown out upon the world, and obliged to accept of a seat in parliament"*.[10]

Eventually the Committee's recommendations, which attempted to reconcile essentially irreconcilable positions and interests, were for a continuance of the licence system, increased surveys by excise officers, higher Lowland taxes, additional taxes on excess production above the limits set by the distilling licence and abolition of the Highland line.[††] This latter was heartily disliked by the excise

---

[††]Excise legislation divided Scotland into three geographical regions, defined by lines: Lowland, Intermediate and Highland for the purposes of taxation. Until its abolition, there was a fourth region, Ferintosh, where no duty was payable. The distinctions, though still referred to, are now archaic.

Figure 2.1   *"Distillers Looking into their Own Business"*, by Thomas
Rowlandson, 1811. The artist was clearly no more impressed with
the London distillers than Commissioner Stodart was with their
Lowland Scottish counterparts.

commissioners, one of whom suggested that a new Hadrian's Wall
with an army of officers to guard it would need to be built to have
any effect on the smuggling of grain into the Highlands and whisky
out of it! Legislation was enacted in 1800 but repeated bans on
distilling over the next decade meant that this was, to a large extent,
a dead letter.

There were further legislative changes, some of which benefitted
the great Lowland distillers who continued their lucrative trade
with England. By now they had acquired an image for sharp
business practice (Figure 2.1) and Excise Commissioner James
Stodart perhaps[‡‡] summed up a widespread view a few years earlier
when he described their *"mysterious and sudden transition from ruin
to pre-eminence and apparent opulency"* going on to consider them
"deservedly odious".

[‡‡]Quoted in David J. Brown's *Politicians, the Revenue Men and the Scots Distillers*, ROSC
12.

In some ways, the story of the next twenty years was written by the atrocious weather that plagued Scotland, with extended and severe winters and poor summers, leading to repeated crop failures, bans on distilling and great hardship and even starvation. Illicit distilling and intimidation of the excise officers seems to have become more widespread. Fearing a major breakdown of law and order, the major landlords reacted to these conditions in a variety of ways. Some pursued a policy of 'improvement' of their estates and there were a number of forced clearances, where people were driven from their homes, most notably by the Duchess of Sutherland and the clan chief MacDonell of Glengarry. Others were more benign in their approach and, where they acted as magistrates, were often reluctant to impose the full penalties set by law, being well aware that it was only through the sale of whisky that rents were paid. A third approach was to promote the establishment of modern, licensed distilleries, such as Brackla (1812), Teaninich (1817), Clynelish (1819) and others now closed.

The foundations for the modern industry had now been laid. Largely under the influence of the poet and writer, Walter Scott, a romanticised national identity was being forged which, by absorbing the (imagined) symbols of Gaeldom and integrating them into a mainstream image of Scotland, simultaneously engaged the Celtic fringe and neutered any remaining threat that it represented to a greater 'North Britain'.§§ It is not for nothing that Scott is referred to as "*the man who invented a nation*"[11] and this was never more overtly expressed than during his management of the state visit to Scotland of King George IV in October 1822, an extraordinary combination of show business, faux ritual and pageantry blended with sufficient *Brigadoon*-style grandiosity to gladden the heart of any Disney 'imaginer'.

The image of plaid, kilt and sporran, foaming cataracts, rocks, mountain and heather and strapping Highland gents in the loyal service of a grateful monarch and Empire would be pressed into service for the next hundred years and more so by an industry that swiftly appreciated the semiotics and potency of national branding. As whisky wrapped itself in tartan, it served to both adopt and

---

§§In Rob Roy (1817), Scott describes "*a Scotch sort of a gentleman... a decentish hallion - a canny North Briton*".

reinforce the iconography of a nation, in a relationship at once synergistic and exploitative.

A later generation of marketers have come to see this imagery as dated and unduly restrictive. Speaking in January 2008, Diageo CEO Paul Walsh said: "*I think the reason that Scotch languished as it did in the 1990s was that, as marketers, we relied too much on the wonders of the product and we communicated those wonders through – forgive me – bagpipes, heather and tartan. Those are very important and relevant qualities but in today's world they are not enough to position the product to a new-age consumer. There comes a point that you are such a believer that you end up talking to yourself rather than ending up talking to a wider audience.*"[12]

From 1820, the state of the industry was earnestly debated and no one was more effective or influential than Alexander Gordon, 4th Duke of Gordon, with his vast Highland estates. The 1816 and 1818 Acts had abolished the Highland Line, reduced spirit duty and the minimum still size and, by permitting thinner washes, assisted the legal Highland distiller. Gordon then argued that further liberal-isation would stimulate the growth of a legitimate, tax-paying in-dustry, alongside which the landed interest would lend their weight and influence to the determined suppression of smuggling.

But Gordon was not alone. Other parties were interested them-selves in the matter. The Rev. L. Moyes, Minister of Forglen (a hamlet between Turriff and Aberchirder) wrote an open letter[13] to the Prime Minister, the Earl of Liverpool, complaining of the vigour with which officers of the excise pursued their duty. "*The poor have their quarters as constantly beat up, their houses entered, their most secret repositories forced open, and their sacred asylum treated with all the indignity of a city taken by storm*". Although Moyes deplored illicit distilling in the most forthright terms, he clearly sympathised with the small Highland distiller attempting to trade legally. In arguing for a reduction in duty, he advanced not just moral considerations but pragmatically demonstrated that revenue would actually increase with a reduced duty if it was ef-fectively enforced.

Moyes suggested that "*by so lowering the duties on the small stills, as to enable the fair dealer to bring his whisky to market, nearly on the same terms with the smuggler: were it nearly on the same terms, the bulk of the people would prefer encouraging the legal distiller*". It was to prove a prescient view.

A Commission of Inquiry into the Revenue was set up but, before it could complete its work, the Illicit Distillation (Scotland) Act was passed in 1822, greatly increasing the penalties for illicit distilling or knowingly permitting it. The flexible powers of local magistrates were revoked, ensuring that a consistent level of penalties was applied, the excise service was given greater authority and its manpower was strengthened. Importantly, restrictions on the export trade were removed, opening English and overseas markets to virtually the entire industry.

In 1823, the following year, the Commission reported and its recommendations were absorbed into the Excise Act, which also introduced the compulsory use of the spirit safe, which had recently been invented by Septimus Fox. The practical effect of the measures was to liberalise the market and assist the small Highland distiller. No longer would the giant Lowland distillers dominate the export business and, equally, the legitimate distillery of quality spirits could compete without handicap with the smuggler.

There was an immediate increase in the number of licensed distilleries, with some 134 new stills commissioned by August 1824 (29 of which were of 500 gallons or more capacity). The impression given in some commentaries on whisky that licensing began in 1823 is false; as we have seen, there was a system of licensed (or 'entered') distilleries long before this, but the significance of the 1822 and 1823 Acts cannot be under-estimated.

Not that these changes were universally welcomed: the Lowland distillers found themselves in some sympathy with the smuggling community, who famously threatened George Smith of Glenlivet, previously a fugitive from the law, and his plan to licence his operation. However, he was not alone and there was a boom in distillery building, some well-planned and financed with a sound understanding of the trade, some not. While most commentators treat the rival activity of smuggling and illicit distillation almost with humour, Michael Given of the University of Glasgow sees in poaching, smuggling and illicit whisky distilling examples of activities which *"directly and explicitly defied the authorities...and allowed people to reaffirm their own agency and self-esteem in the face of humiliation, eviction, exploitation and what they saw as the violation of the social contract."*[14]

Between 1824 and 1829, as well as Glenlivet, we can record the establishment (or re-opening as legitimate) of Cardhu, Fettercairn,

Macallan, Ben Nevis, Bowmore, Highland Park, Glendronach, Pulteney, Laphroaig, Aberlour, Miltonduff and Springbank, as well as many others which, like Port Ellen, Islay (which operated in 1825 as a mill, but was distilling by 1833), have subsequently fallen silent. Legal production grew from 3 million to over 10 million gallons during the same period. Along with Cameronbridge, a further 15 Lowland distilleries were licensed.

But in the Lowlands something else was stirring. John Haig began testing an experimental still at Cameronbridge in 1827 to a design by Robert Stein, who had also begun trials in his brother's Kirkliston distillery in 1829. This was the beginning of the continuous still, which was to transform the industry. By the end of 1830, Haig's still at Cameronbridge had produced close to 150 000 gallons of whisky – around thirty times what a good-sized Highland pot still distillery could make in the same time.

What is more, the spirit produced was palatable, though bland, and high in strength, making it ideal for rectification. However, virtually simultaneously, in Dublin, Aeneas Coffey – a former Inspector General of the Irish Excise turned distiller – was working on his own design, which he patented in February 1831. Although some Stein Stills were to continue operating until the 1880s, the Coffey design was the superior by virtue of its greater strength, simplicity, inherent safety and ease of operation. It had a capacity of some 3000 gallons of wash per hour and produced spirit at 94–96% alcohol by volume (abv). Within three years, Coffey had sold seven of his stills to Irish distillers and was ready to conquer new markets – his first Scottish customer was Philp of Alloa. Further plants were installed in rapid succession so that by 1836 Coffey Stills were making around 30% of Scotland's grain whisky. Although expensive, the new column stills were efficient and made large quantities of high-quality spirit at a competitive price.

Ironically, however, the pioneers over-extended themselves, running up large debts and being forced into sequestration – Andrew Stein in 1831, Andrew Philp in 1834 and William and Robert Haig in 1835. Despite this, the upward march of the Coffey Still continued and a manufacturing plant was established in London in January 1835. As well as supplying the UK market, Coffey Stills were exported, especially to Australia and the West Indies.

It would seem that rather too many distilleries were opened optimistically at this time and many soon failed due to a number of factors, such as poor planning and lack of finance, failed harvests, economic recession, continued competition from smuggling, and by 1844, sixty or more firms had failed. That said, a number of new entrepreneurs entered the trade, initially as grocers or wine and spirit merchants, and their fledgling businesses were to prove enormously important. At the same time, the social climate in Scotland was changing, with greater attention being paid to the dangers of excessive drinking and the temperance and prohibitionist cause, often led by clerics, gathered momentum.

According to Moss and Hume's account of the Scotch whisky industry,[3] the scale of the largest distilling operations *"far surpassed the largest of the new iron works and marine-engine works"* (as the Stein works at Kennetpans and Kilbagie had done previously) but this was not widely acknowledged at the time. History has thus tended to ignore the significance of distilling in the early industrialisation of Scotland; the continued deterioration in trade conditions and further distillery closures up until the late 1850s hardly helped matters.

And then a series of happy coincidences combined to transform the fortunes and status of Scotch whisky over the next 30–40 years. In summary, the UK economy began to recover and then expand; transportation became significantly easier due to the growth of the railways which, as well as more efficiently carrying supplies and finished goods, brought tourists to an increasingly fashionable Scotland; Scottish regiments, administrators and traders spread out across a growing British Empire; new legislation assisted the increasingly common practice of blending and, most fortuitously of all, a small aphid began consuming the vineyards of Europe.[¶¶] As brandy production slumped, something was required to step into the gap and a new generation of entrepreneurial distillers, blenders and merchants ensured that it would be Scotch whisky.

The practice of 'vatting' appears to have been initiated by Andrew Usher, an Edinburgh whisky merchant, distiller and agent for The Glenlivet. His 'Old Vatted Glenlivet' dates from 1853; other early pioneers include W. P. Lowrie of Glasgow, Charles Mackinlay, W. A. Robertson (later of Robertson & Baxter) and

---

[¶¶] The first vines infected with *phylloxera* were inadvertently planted by a M. Borty in his vineyard in Roquemaure in the Gard in the spring of 1862. The Scotch whisky industry might consider erecting a plaque here.

William Sanderson of Leith. Blending took some while to settle down: early practice, which in places survived into the 12th century, was not averse to adding a cocktail of ingredients. One coldstream merchant, not untypically, happily added rum or sherry 'for colour' while Sanderson's recipe books even list a cherry whisky!

Home markets in both England and Scotland were dominated by Irish whiskey at this stage, largely due to its greater consistency. *"Whisky was something barbarous from across the border"*, writes Alec Waugh in his history of the Gilbey company[15] noting that *"in 1875 Gilbey's sold 83 000 dozen Irish [bottles] to only 38 000 dozen Scotch"*. Blending was to prove the solution and the Scottish distillers took to it with a will, although an Irish style of spirit was produced, especially at Edinburgh's Caledonian distillery.

The later campaign by the great Dublin distillers alleged the adulteration of Scotch whisky with 'Hamburg sherry', 'prune wine' and 'cocked hat spirit' – apparently, with some justice and Edward Burns[16] has documented widespread adulteration in Glasgow in the 1860s and 1870s. Eventually, this was countered by the introduction of branded containers, both stoneware flagons and the first bottles. The distillers became increasingly anxious to protect their brands – understandably so when the whisky carried the family name.

Blending took hold with astonishing rapidity. The great houses of Walker, Dewar, Buchanan, Gloag, Bell, Chivas, Teacher and Ballantine all came into their own, along with countless others. Starting as general merchants, tea blenders or 'Italian warehousemen', they began to specialise in the new blended whiskies although, to the relief of the pot still distilleries, Irish whiskey was still a huge seller in Scotland. The English trade developed relatively slowly – Winston Churchill, speaking of this period, said: *"my father could never have drunk whisky, except when shooting on a moor or in some very dull chilly place. He lived in the age of brandy and soda"*.[17] A brand, such as Gloag's Grouse whisky (today, The Famous Grouse), was built on such outdoor, sporting associations.

There was a remarkable burst of construction as new distilleries were built, many designed or improved by the noted Elgin-based architect, Charles Cree Doig, accompanied by growing financial speculation (made possible by easy credit terms offered by many Scottish banks) that reached fever pitch in the 1890s. At the same time, a number of the blending houses, many by then in the hands of a remarkable group of second generation sons, attacked the London market with some determination – eventually three firms came to a position of particular prominence: John Walker & Sons (founded in Kilmarnock, 1820), John Dewar & Sons (Perth, 1846) and James Buchanan & Co. (London, 1884). Later, John Haig & Co. and Mackie & Co. would join them and they collectively became known as the 'Big Five'.

During the same period, the grain distillers grouped together, first by one-off agreements in 1857 and subsequently in 1865–1868, leading eventually to the creation of The Distillers Company (DCL) in 1877. This comprised the grain whisky producers of Cambus, Cameronbridge, Carsebridge, Glenochil, Kirkliston and Port Dundas with the Caledonian Distillery (Edinburgh) negotiating an arms-length agreement. By 1883 the DCL was a publicly quoted company and the Caledonian Distillery had joined the grouping.

The archetypal image of Scotland, leaning heavily on the fantasies of Sir Walter Scott, was pressed into service at this time, along with effusive endorsements by medical practitioners as to the health-giving qualities and purity of the brand. There was intense competition at the many international trade fairs that followed the success of the Great Exhibition of 1851 (at which Aeneas Coffey had exhibited his patent stills, receiving a medal 'for services' from Prince Albert). The medals that were awarded were quickly incorporated into label designs and advertising and were seen as a powerful endorsement.

The distillers and blenders advertised and promoted their brands lavishly, none more so than the Leith firm of Pattison, Elder & Co. and the building boom continued, with sixteen new distilleries constructed between 1895 and 1900. All too often these were speculations, the promoters hoping to quickly sell out once the new distillery was in production. At the same time, stocks of whisky began to grow rapidly, egged on by eager bankers who appeared happy to lend against stocks of maturing whisky. A bubble began to form, with Pattison's at its heart.

The firm was run by two brothers, Robert and Walter Pattison, who had inherited partnerships in a small and entirely respectable brewery and a whisky blending business in Leith. In 1896, they floated it as a company, taking all of the ordinary shares, a quarter of the preference shares and £150,000 in cash as payment. Just as in our own dotcom boom, the share offer was six times over-subscribed and, just as in the dotcom boom, the directors could initially do no wrong.

"The Doctor", 'Morning Gallop' and 'Morning Dew' brands were blends, but Pattison's also owned distilleries at Aultmore and Glenfarclas (in partnership with the founding family) and substantial office and blending premises in Leith and London.

They were noted for the extravagance of their personal lives and for their heavy advertising (Figure 2.2). One of their promotional schemes, in particular, was a stroke of pure genius: the Pattisons purchased five hundred African Grey parrots and gave them to publicans and licensed grocers. However, their proud owners soon discovered they'd been trained to squawk *"Pattison's whisky is best!"* and *"Buy Pattison's whisky!"* at unsuspecting customers.

Figure 2.2   Pattison's advertised heavily – and perhaps prophetically.

By December 1898, the DCL, who had continued to extend credit to the brothers, was now owed over £30,000 (*i.e.* more than £2m) and the directors became alarmed. The Pattison's account was frozen. In the same month, an attempt to sell some stocks quickly and cheaply spread alarm in the trade. The Pattisons tried to draw around £20,000 from their bank but, with less than half that in their account, the result was predictable. Their collapse followed swiftly and was initially greeted with incredulity. It seemed scarcely possible that a firm of their size and reputation could just go under. Had Alfred Barnard himself not described them[18] as "*one of the most thriving and go-ahead concerns in Scotland... its rapid success is largely due to its present enterprising proprietors*".

Attempts were made to mount a rescue early in 1899 but slowly, the truth emerged: the Pattisons had massively over-valued their stock and inflated their profits, often by the simple device of selling the same whisky (on paper) several times over.

Moreover, they appeared to possess a magical touch with the blending vat, claiming to produce well-matured blends with a high proportion of malt. The sinister truth was that this was a further manipulation. As the *Wine Trade Review* reported:

> "*The Pattisons purchased an enormous quantity of new grain whisky at $11\frac{1}{2}d$ per gallon. They put it through a vat, giving it a dash of malt and it suddenly came out as good Glenlivet, worth 8/6d per gallon. The total cost of the whisky in this 'Glenlivet vat' was 2s $3\frac{1}{2}d$ per gallon. Three quarters of that was cheap grain whisky".*

At their resulting trial, the jury took just an hour and a half to find the Pattisons guilty of fraud, Robert being jailed for 18 months and Walter for 9 months.

More than the Pattisons and their creditors suffered. Ten other firms were brought down in the collapse that followed. The reputation of blending, and indeed of the whisky trade in general, was hard hit, not recovering until the Royal Commission of 1908/1909 – only to be hit once again with punitive tax increases in Lloyd George's budget of 1909.

But others benefited, most notably, the DCL. Their financial prudence throughout the boom years stood them in good stead as trade tightened; they were able to purchase a number of the Pattisons' assets at knock-down prices and accordingly they steadily came to assume a position of leadership in the whisky business. In later years, William Ross (Figure 2.3) of the DCL wrote,[19] somewhat sanctimoniously, that "*...their extravagance in conducting business, including the somewhat palatial premises, was the talk of the Trade, but so large were their transactions and so wide their ramifications that they infused into the trade a reckless disregard of the most elementary rules of sound business*".

Ross was being more than a little disingenuous, however. As one of the principal suppliers to Pattison's, he had dangerously over-extended their lines of credit and must carry some responsibility for not acting earlier on his own well-founded concerns.

Perhaps the last word should rest with the liquidator of Pattison's who, in his final report, described these events as constituting "*the most discreditable chapter in the history of the whisky trade*". Pattison's were Scotland's equivalent of an Enron scandal.

Figure 2.3 William Ross.

Ross may have spoken of "*speculators of the worst kind drawn into the vortex [who] vied with each other in their race for riches*" but, with hindsight, there is a kind of inevitability to the whole affair. If it had not been for the Pattisons then, in all likelihood, another firm would have failed, exposing the whole fragile edifice of an over-blown trade driven by speculation, greed and chronic over-production: so much for the image of the canny and parsimonious Scot.

Over-production of whisky had reached spectacular levels in a market about to be hit by an economic slump. But there was to be one beneficiary. The DCL moved to acquire a number of their competitors, some more than willing to exit the business to a ready buyer and, by closing distilleries and reducing capacity, they thus removed much unprofitable excess supply from the market. Ross proved a determined, indefatigable and clear-sighted leader. Such a man was shortly to be needed.

The question of what exactly constituted whisky had troubled commentators for some years. There was extensive correspondence in the press, notably the *Daily Telegraph*; the leading Irish distillers had campaigned volubly against the use of 'sham spirit' (grain whisky), as later did a hard-core of backwoodsmen within the North of Scotland Malt Distillers' Association (NSMDA) and the medical journal *The Lancet* interested itself in the case. Claims and counter-claims, accusations and increasingly hysterical assertions were flung about with abandon. "*...this villainous 'blend' is responsible for the majority of lunatics who crowd our asylums and infirmaries. I have known cases where young, healthy, strong men have become raging lunatics after partaking of a few glasses of it*" stated one correspondent,[20] signing himself "*Police-Inspector of 26 years' experience*".

A pamphlet, titled "*The Practice of Substitution in the Spirit Trade*" was circulated, reprinting a long article from *The Lancet* criticising the indiscriminate use of silent spirit and discussion in the press continued. The following year (in 1904) Sir Walter Gilbey published his *Notes on Alcohol,* in which he argued strongly for the direct-fired pot still. "*When a plain, flavourless spirit, the produce of the patent still is blended with a pot still whisky*", he wrote "*it is but a very poor compromise...*".

Faced with increasing stocks and low levels of trade, the NSMDA was split – some members favoured an active campaign to promote malt whiskies, while others, conscious that the blenders were becoming ever more important customers, recommended restraint and compromise. Their case was thus weakened but initially appeared to have the upper hand, both in the press and in law. Efforts were made in the House of Commons to introduce legislation defining whisky in favour of that made in a pot still, but little progress was made.

In November 1905, a case was begun by Islington Borough Council (according to R. B. Weir:[21] *"a Labour controlled council with strong temperance sympathies"*), who had previously brought prosecutions for the adulteration of brandy, against two publicans under Section 6 of the Food and Drugs Act, in which it was argued that the purchaser had received an *"article...not of the nature, substance and quality demanded"*. Asked for Irish and Scotch whisky, the accused sold blends – in the case of Scotch whisky, a blend comprising 90% grain and, perhaps unsurprisingly, the magistrate found against them. This was a clear vindication of the malt distillers' more extreme position. Some, like Powers' in Ireland, rushed to celebrate, issuing postcards marking *"another Irish victory"* (Figure 2.4).

Figure 2.4   A premature victory celebration by Powers' of Dublin in the 'What is Whisky?' case, 1906.

However, that victory was a hollow and temporary one. A battle had been won, but not the whole war and the grain distilling fraternity, led by the DCL, returned to the fight.

The verdict was appealed but the magistrates returned a split verdict, leaving the original decision in place. Further representations were made and, after some lobbying, a Royal Commission was set up in February 1908 to provide a definitive answer. The DCL promptly began advertising their 7 year old Cambus grain whisky under the splendid slogan: "*not a headache in a gallon*" (Figure 2.5).

However, if the industry had been hoping for an absolute ruling, they were to be disappointed. The Royal Commission reported in July 1909 and avoided any verdict on the percentage of malt whisky to be contained in a blend and was also silent on the vexed

Figure 2.5   Cambus grain whisky was heavily advertised as part of the DCL's campaign to gain respectability for the product of the patent still.

issue of a minimum maturation period for whisky. It did, however, accept that patent still spirit distilled in Scotland could be termed 'whisky' (the report in fact referred to 'whiskey'); thus the Royal Commission turned out to be a significant victory for the blending industry.

Ross of the DCL had led much of that campaign, but was promptly faced with another stern challenge from the Chancellor Lloyd George's People's Budget. Seeking to fund old age pensions and unemployment benefit, David Lloyd George targeted distillers, increasing duty by a punitive one-third, while leaving duties on wines and beer at existing levels. This was no accident. He was an ardent teetotaller and saw it as his moral duty[||] to help drive the Scottish people "*by tax on whisky to lighter beer*".

In this, he was to some degree successful. Despite substantial increases in tax rates the actual revenues fell as sales dropped. Very soon there was a wave of distillery closures and company failures that continued until the advent of the World War I, when there was a short-lived recovery.

By 1915, however, Lloyd George returned to the attack, blaming the drinks industry for the poor productivity of the British armaments industry. A shortage of high explosive in France led to the so-called Shell Crisis, a significant factor in the fall of the Liberal Government, in favour of a coalition. Lloyd George's rise to power as the new Minister of Munitions continued; eventually he would replace Asquith as Prime Minister in a further political crisis in December 1916. Buchanan's and Dewar's merged in 1915, an earlier attempt to arrange a larger merger involving Walker's having foundered.

Lloyd George's first proposal was for total prohibition for the duration of the war and, as a step to this, he sought a further doubling of the tax on spirits; fortunately for a beleaguered industry this proved unacceptable to Irish Nationalist MPs and the distilling industry came forward with a proposal for a minimum age of maturation for whisky. This was accepted but, while it may actually have favoured those producing a higher quality blended whisky, it was combined with further measures controlling the drinks trade and the retailing of alcohol. More distilleries fell silent.

[||]Concerned as he was with others' morals, Lloyd George was a notorious philanderer. His nickname 'The Goat' testifies to his sexual appetite.

By the end of World War I, both production and consumption of whisky had fallen drastically, although the patent still distilleries did not suffer so badly as they were able to convert their plants to make materials needed for the war effort. The hoped-for recovery that might have been expected at the end of the war did not materialise as the 1919 Budget drastically increased taxation and restricted the trade from passing on the full increase to the home market consumer.

Further closures and mergers followed, increasing the strength of the DCL which was becoming ever more dominant. Sales continued to decline, partly due to tax increases in the 1920 Budget and the arrival of prohibition in the USA, which had become increasingly important. Australia, then Scotch whisky's largest export market, and Canada soon introduced tariff barriers to protect their own producers and further consolidation was to follow.

Possibly the most important step was the merger in 1925 of Walker, Buchanan – Dewar and the DCL under the chairmanship of William Ross. The flamboyant Sir Thomas 'Tommy' Dewar greeted the union in typically florid style, describing Ross[17] as *"our new Moses to take us out of the wilderness of strenuous competition and lead us into the land flowing with respectable dividends"*. Quite what the ascetic Ross made of this is not recorded.

The recessionary climate continued. 1926 was, of course, the year of the General Strike so Scotch whisky was not facing hardship in isolation but the statistics are stark: around 50 distilleries had closed since 1921; only 113 were working in 1926 and that fell to 84 the following year. This did not go unnoticed to contemporary commentators. In *Caledonia.*[22] the nationalist writer and journalist, George Malcolm Thomson,[***] noted that *"the whisky industry is in an even worse plight, as a result of high taxation and American prohibition. In one important centre only one distillery out of seven is working"*.

It should not be imagined that the Scotch whisky industry simply sat back and watched the important US market simply disappear due to the success of the abolitionists. On the contrary, many in the trade made vigorous, if covert, efforts to maintain their business.

[***]Three years later, under the pseudonym Aeneas MacDonald, he was to write his best-known work, *Whisky*, the first modern book on the subject.

In Canada, Bermuda, Cuba, British Honduras, St Pierre and Miquelon and, above all, in the Bahamas, a number of agents were appointed to import whisky entirely legally. But the quantities involved were remarkable: in the Bahamas, for example, which grew rich on the trade, imports of less than 1000 gallons of Scotch in 1918 had grown to close to 400 000 gallons in just four years. The 6000 inhabitants of St Pierre and Miquelon were apparently consuming 20 gallons each annually, children included. (In fairness, similar statistics could be produced for gin, rum, brandy and wine.)

A mixed band of real sailors, drifters, adventurers and plain gangsters assembled a veritable armada of ships to take this whisky to the waters off the New York/New Jersey coast, just outside the 12 mile limit, which became known as 'Rum Row' (something similar also happened on America's west coast). There it was transferred to speedboats, fast launches and, on occasion, flying boats, all of which could out-run the slower coast guard vessels. By such imaginatively clandestine means, the eagerly sought-after bottles were delivered to the thirsty customers of America's 'speakeasies'.

Despite vigorous efforts by the authorities, the dangers of the trade, the possibility of a cargo being hijacked at gunpoint by unscrupulous gangsters and the risk of imprisonment, the illegal business flourished. The distilling industry was very well aware of what was going on – questioned in 1931 by a Royal Commission, Sir Alexander Walker of John Walker & Sons was adamant that the trade would continue unabated. Respectability was maintained by the polite fiction of shipping the whisky to legal and legitimate agents – what the agents then did with the whisky was their business.

However, while some distillers prospered and suppliers to the smuggling business, such as Samuel Bronfman in Canada, eventually parlayed their way to respectability, for others it was disastrous. The Irish whiskey industry never really recovered and some commentators suggest that the willingness of the Campbeltown distillers to supply substantial quantities of whisky with little regard to quality was a factor in their demise after legal trade was resumed, although it seems equally probable that the loss of cheap supplies from an exhausted local coal field may have been as important.

In 1935, after a long and distinguished career of 57 years during which he was blinded in an accident while on company business,

William Ross retired as Chairman of the Distillers Company. He was a central and highly influential figure in the Scotch whisky industry and, in many ways, responsible for the structure of the modern industry. His political skills were seen at their best during the early years of the century, culminating in victory for the blending interest in 1908/1909 and again in the early 1920s leading to the merger of the 'Big Three', moves which remain of enduring importance. During his period running the company, the DCL had many interests beyond whisky distilling and Ross shrewdly used the strength this gave the company for a position of great influence and authority, increased by his personal reputation for integrity.

In the years prior to the start of World War II a slow recovery and a gradual consolidation of ownership was seen, which accelerated after the war into international rather than Scottish hands. Working with the National Distillers of America, Joseph Hobbs, a Scottish – Canadian businessman, acquired eight distilleries by 1938 and Hiram Walker purchased Glenburgie and Miltonduff distilleries, the blenders, George Ballantine & Son, and built a giant distilling, blending and bottling plant at Dumbarton.

Once again, wartime saw an increase in taxation and restrictions on distilling as grain was required for food. By late 1943, every malt distillery in Scotland was closed (while grain distilleries remained open to produce ethanol, which was vital to the war effort) and prices were soaring. Priority was given to exports, especially to the USA, which became even more important to the country's financial position once the war was over. The war years also saw the formation of the Scotch Whisky Association (SWA) in something approaching its contemporary form. That body was influential in restoring price stability to the UK market by buying up supplies and by restricting availability to retailers and wholesalers who were seen to exploit the shortages by inflating prices.

Between 1945 and 1953, the industry was hedged about with restrictions on the quantity of whisky which could be distilled, further tax increases, strict limits on sales in the home market and continued pressure to increase export sales. Hard though it is to believe at this juncture, around half of all exports went to one single market, the USA, where Scotch enjoyed an enviable market position.

However, following the removal of government controls in 1954 the industry went into a period of sustained growth. New distilleries were built and existing ones rapidly expanded. The overwhelming majority of sales were in standard blended whiskies; there was relatively little stock available for premium expressions and, outside of Scotland, 'self' or single malt whiskies were little known. In 1963, for example, Harpers Directory listed more than 2000 Scotch whisky brand names of which only 33 were for single malts. Many were hard to find or available only by application to the distillery.

Though available in limited quantities well before World War II, active post-war marketing of single malt (or 'pure malt' as it was styled) began in 1964 when William Grant & Sons launched Glenfiddich, partially in reaction to the aggressive tactics of the Distillers Company following Grants' proposal to commence TV advertising for their Standfast blend (a gentleman's agreement amongst the industry had decided to eschew the new-fangled commercial television channels, first launched in 1955). Grant also embarked on the ambitious construction of a new grain whisky distillery at Girvan to ensure their own future supplies. Remarkably, this was built in around nine months, on the site of a former munitions factory.

Exports, particularly to the USA, continued to grow rapidly through the 1960s when Cutty Sark, J & B Rare and Johnnie Walker enjoyed great success there. In 1962,[†††] Cutty Sark sold over 1 million cases in the USA alone – over the next ten years this doubled. Its competitors achieved similar growth.

As late as 1969 the Glasgow stockbrokers, Speirs & Jeffrey, were confident in their view of the future.[23] "*We do not foresee any slackening in the long term growth in demand for Scotch whisky: increases will vary from year to year, but we expect that an 8–10% per annum growth in exports will be achieved*". They went on to comment that "*there is still an enormous stock of whisky over-hanging the market...[but] we think... that the bogey of over-production has been laid to rest convincingly*". Their optimism was to prove sadly misplaced. As a nostalgic commentary on the industry it may also be noted that Speirs & Jeffrey were then able to

---

[†††]Interestingly, this was achieved before the widespread adoption of containerised shipping, meaning that each and every case would at some stage have been moved by hand.Losses of whisky, whether from breakage or pilferage, were dramatically reduced with the introduction of the steel shipping container.

comment on the prospects of fourteen separate quoted companies. By contrast, today's *Scotch Whisky Review*[24] analyses just two.

Because of this steady growth and perhaps also because of the greater number of competing companies (and there were many more privately-owned distilleries at this period also) this was something of a halcyon period for the whisky broker. The hard-drinking, Jaguar-driving, golf-playing whisky broker of legend made his living – often a handsome one – by trading parcels of whisky, sometimes on commission, sometimes on his own account, in essence using his contacts and deep knowledge of the industry as arbitrageur. Some, such as Stanley P. Morrison, made the journey to respectability as a distiller in their own right but, for the most part, this colourful character has been squeezed out of the modern industry.

The long period of growth slowed briefly towards the end of the 1960s but then picked up pace with further distillery openings and expansions. Whisky was now enjoying success in many markets other than the US – France and Japan in particular – and exports continued growing until 1978. There was yet more consolidation and change of ownership – 1972 was notable for the purchase of the Aberlour distillery by Pernod Ricard of France, who were to become a very significant part of the industry in the future. Many of these changes were later to be undone or subject to further change as the ownership of the Scotch whisky industry went through a period described[3] as *"almost a paradigm of modern capitalism: take-over, merger, demerger, management buyout, globalisation and deglobalisation are all there"*.

During this period, the extraordinary position of dominance achieved by the DCL was slowly fading. The group operated as separate companies but despite this inherent inefficiency still had a global market share approaching 50% of all Scotch whisky as late as 1973. But problems, for them and for the industry were fast approaching.

The first difficulties arose as a result of the Arab–Israeli War in October 1973. The dramatic rise in oil prices that followed led to a widespread slowdown. To compound the problem, drinkers in the USA switched in considerable numbers to vodka and white rum, while the ending of the Vietnam War following the fall of Saigon (now Ho Chi Minh City) in April 1975 led to further disruption in the American economy, previously boosted by government spending.

The DCL were then hit by an EEC Commission ruling banning dual pricing, a system employed to generate margins in export markets which local agents used to fund promotional activity. Though not alone in employing this tactic it was of considerable importance to the DCL. Seeking to protect their European agents, some brands were withdrawn from the UK (Johnnie Walker Red Label being the most prominent) and prices were raised. This led to a large loss of market share and an impression of vulnerability in the company's hitherto impregnable image.

As a result of the downturn, there were further rationalisations and acquisitions, often led by the UK's large brewing groups, then protected by the tied house system.[‡‡‡] As one example, International Distillers & Vintners (IDV), owners of the powerful J & B brand, were acquired by the brewers, Watney Mann, but they were themselves taken over by Grand Metropolitan. In 1975, Long John International and, with it, D. Johnston & Co., proprietors of the Laphroaig distillery on Islay, were acquired by Whitbread & Co. Ltd; however, they subsequently sold the companies and, in 2001, completely withdrew from the alcohol business (after 259 years continuous operation as a brewer).

A number of companies, such as Matthew Gloag & Son, Teacher's and Whyte & Mackay, lost their independence, while smaller concerns began to look to the single malt market to maintain theirs. To make matters worse, there were five consecutive duty increases on spirits between 1979 and 1985, the government appearing to care little for the state of the industry.

There was some positive news in 1974 with the creation of the Pentlands Scotch Whisky Research laboratories (based on the consultancy expertise of the Arthur D. Little firm), which aimed to provide the independent distilling industry with scientific, analytical and research and development services to compete with the in-house facilities of the DCL and Chivas Brothers. Pentlands did much work on flavour-related and spirit yield issues but is best-

---

[‡‡‡]Prior to the 1989 Beer Orders, UK brewery companies were permitted to 'tie' an unlimited number of on-trade outlets, requiring them to buy most or all of their supplies from the brewery. Consolidation of the brewing industry after World War II led to the creation of very large retail estates controlled by a small number of companies. Seeking to open up the market to greater competition, the 1989 Act restricted the tie to a maximum of 2000 public houses but inadvertently led to the creation of property-owning 'pub companies' – in effect, the tie but by another name.

known, at least to consumers, for their pioneering work on the 'whisky wheel', starting around 1979.

This was the first systematic attempt to define the language of whisky tasting: now the accepted way of tabulating aromas and flavours, at the time it was novel. Initially, the wheel was intended for industry use only but, with increasing consumer interest and growing connoisseurship, a more extensive consumer version was developed in the mid-1980s. (See also chapter 5, where the whisky wheel is discussed in greater detail.)

Recognising the need to expand Pentlands' work, it later grew into the Scotch Whisky Research Institute, established in October 1995, and is now located in purpose-built facilities adjacent to Heriot – Watt University.

However, if there were not troubles enough back in the 1970s, to make matters worse, further dark clouds were gathering as the 'whisky loch' of ill repute gradually filled up with unwanted Scotch whisky. There had been steady over-production of both malt and grain whisky during the late 1970s and early 1980s. Younger consumers, especially in America, were continuing to turn to vodka and white rum and sales of all brown spirits were falling. In the home market, following the UK's entry to the Common Market in 1973, the tax system increasingly favoured wine at the expense of beer and spirits. For some years, a complacent industry, largely production-driven, ignored this in the belief that the effect was temporary and that drinkers would return to whisky after a brief flirtation with other spirits. Many years of continuous growth had lulled the Scotch whisky industry into a comforting sense of optimism that was to prove sadly misplaced.

Ardbeg was closed in 1981, a relatively modest first step compared to what would shortly follow. In 1983, the DCL closed Carsebridge grain distillery, 11 of its 45 operational malt distilleries and two bottling halls, following this up with a further round of closures in March 1985, when a further ten single malt distilleries were closed. Other distillers took similar action: Bunnahabhain was mothballed and the other Highland distillery plants were placed on short-time working; Tomatin Distillers went into liquidation; Glen Flagler (Airdrie) was closed and Glenglassaugh and Ben Nevis were mothballed – by 1986, around a quarter of all Scotch whisky distilleries had gone out of production. In 1983, Glenfarclas worked at around 50% capacity; however, flouting the trend, this family-owned distillery upped production the following year reasoning, correctly as it turned out, that there would soon be a shortage of 3–4 year old malt fillings.

UK shoppers enjoyed a plethora of cheap own-label brands (some containing whisky of surprising quality) and price promotions as distillers struggled to sell unwanted stocks – an unintended consumer benefit of the whisky loch. The most dramatic moves, however, came between 1984 and 1986 with the acquisition of Bell's of Perth by Guinness PLC and Guinness' subsequent move to take over the DCL in a bid contested by the supermarket Argyll Group.

This was a hostile, bitter and hard-fought business; "*a campaign of unprecedented viciousness between Argyll and Guinness*",[25] in which reputations were tarnished and the public image of the drinks industry (not to mention the City) brought into disrepute. Later, key participants were to serve jail terms for offences related to the manipulation of the Guinness share price, including false accounting, conspiracy and theft. The final result, however, was the creation of United Distillers in 1987, representing the merger of Arthur Bell and the DCL, although regulatory pressure forced the sale of a number of brands to Whyte & Mackay of Glasgow.

At this point, the DCL was still operated as a loose grouping of nearly 40 subsidiary companies reflecting the original businesses, large and small, which had made up the group. Maintaining their own head offices (some quite ostentatious), infrastructure and management effectively, they competed in the market against their sister companies on all critical measures, except price. James Espey, a former senior sales and marketing manager and chairman of Seagrams described the DCL companies as resembling "*an uneasy confederation of fiefdoms, each one at odds with the others*".[26]

There were few, if any, integrated global marketing strategies and international distribution remained largely in the hands of independent third-party agencies, making it well-nigh impossible to impose any degree of central control or consistency of brand message. In effect, the Scotch whisky industry was hamstrung by its own structural inefficiencies and the associated cost penalties of what had become an archaic and inefficient way of working.

That this had been recognised by outside forces was a key factor in the Guinness takeover of the DCL. Ernest Saunders, the dynamic M. D. of Guinness, saw the opportunity very clearly and, once in control of the company, began a ruthless if necessary and long-overdue process of consolidation of the various DCL companies under one corporate umbrella. The logic was unarguable, even if the short-term consequences were traumatic for many employees, and his work was continued by Guinness' new group chief executive, Anthony Tennant, after Saunders' eventual disgrace and imprisonment.

As S. R. H. Jones of the University of Dundee has argued,[25] another formative and highly influential factor in the development of Scotch whisky at this time was rapidly growing sophistication in the marketing function; in particular, the systematic application of lifestyle segmentation in increasingly effective brand management strategies. In fact, it is no exaggeration to suggest that the 1980s were the decade in which Scotch whisky evolved, painfully in some cases, from a production- to a marketing-led industry.

Ten years later, United Distillers & Vintners (UDV) was created by the integration of United Distillers and Grand Metropolitan's IDV and, shortly afterwards, Guinness and Grand Metropolitan merged to create Diageo (1997). In 2001, Diageo acquired much of the Seagram's wine and spirit interests in a joint acquisition with Pernod Ricard, Seagram having decided to exit the drinks business in favour of media and entertainment markets.

Today, Pernod Ricard is the significant second force in Scotch whisky (behind Diageo), controlling major brands, such as Chivas Regal, Ballantine's and Royal Salute, all of which are particularly important in Far Eastern markets, and operating some 14 distilleries. Their single malt brands include The Glenlivet, Aberlour, Strathisla and Longmorn.

During this decade there were many other varied disposals and transfers of ownership, notably the creation of an enlarged Inver House Distillers, who acquired Knockdhu and Speyburn from UD, these being the first disposals of old DCL distilleries. This was shortly followed by the sale of Imperial and Glentauchers, but the most significant change was the major disposal of the Dewar's brands and four distilleries as a condition by the regulatory bodies to permit the creation of Diageo.

Dewar's was bought by the Bacardi Corporation who, until then, had been a relatively minor presence in the Scotch whisky industry owning only the MacDuff distillery and the William Lawson's blend. They have subsequently expanded production at their sites and extended the Dewar's brand range.

Whyte & Mackay's market presence was greatly enhanced by the purchase in 1986 of a number of brands from DCL, some on a world-wide basis and some for the UK only. The company has been through a considerable amount of structural change, however, and over a period of less than 40 years has been part of Scottish & Universal Investments, Lonrho, Brent Walker, Gallaher (American Brands, subsequently Fortune Brands, during which time they purchased the Invergordon grain distillery), Jim Beam Brands, a management buy-out (2001) briefly named Kyndal, followed by acquisition by South African entrepreneur, Vivian Immerman, who in May 2007 sold the business to Dr Vijay Mallya's United Spirits of India. More recently, Diageo have acquired significant interests in and effective managerial control of United Spirits and the structural implications for Whyte & Mackay remain unclear at the time of writing.

Other important groupings include The Edrington Group, LVMH (Glenmorangie PLC), La Martiniquaise, William Grant & Sons and Loch Lomond Distillers.

The Edrington Group (which is privately held by a charitable body, The Robertson Trust) owns Highland Distillers and with it important brands, such as The Famous Grouse, The Macallan and Highland Park, as well as a joint venture with William Grant & Sons (The 1887 Company). In recent years, Edrington has pursued a policy of concentrating resources on their principal brands, disposing of smaller or less well-known distilleries, such as Bunnahabhain (to CL World Brands' operating company, Burn Stewart), Glengoyne (to Ian Macleod Distillers), The Glenrothes (to Berry Bros & Rudd), Glenglassaugh (to the Scaent Group, who subsequently re-sold to BenRiach Distillery Company) and Tamdhu (also to Ian Macleod Distillers). Edrington purchased the Cutty Sark brand from Berry Bros & Rudd and have also diversified out of Scotch whisky with the purchase of a majority stake in the Dominican Republic's Brugal rum producer.

Louis Vuitton Moet Hennessy (LVMH, owners of Glenmorangie) and La Martiniquause are both − like Pernod Ricard and Remy − Cointreau (owners of the Bruichladdich single malt) − French but, otherwise, have relatively little in common, operating as they do at opposite ends of the whisky market. In broad terms, LVMH is concerned with premium luxury brands, where La Martiniquause concentrates on value offerings.

LVMH acquired Glenmorangie PLC in late 2004 and has concentrated on repositioning the company to operate exclusively in premium sectors. Accordingly, the company relinquished much supermarket own-label business, reduced its exposure to blended Scotch and sold the Glen Moray distillery and brand to La Martiniquaise. Glenmorangie also owns the Scotch Malt Whisky Society, a membership club for single malt whisky enthusiasts, but this is operated at arm's length. The company's principal brands are Glenmorangie and Ardbeg, an Islay distillery with an enthusiastic cult following.

La Martiniquaise operates the Glen Moray distillery and a substantial, if little known, modern grain distillery and bottling plant at Starlaw, near Bathgate. The Starlaw distillery has a capacity in excess of 25m ola (original litres of alcohol) annually and there are also plans to develop a large single malt distillery on an adjacent site. Presumably the bulk of this output will be utilized in the company's Label 5 blend, which sells in excess of 2.5m cases largely, but not exclusively, in France. The company is principally owned by the Cayard family. Its other main whisky brand is the Glen Turner malt.

William Grant & Sons are probably best known for the pioneering and highly successful development of their Glenfiddich single malt, but also own The Balvenie, Kininvie and Ailsa Bay single malt distilleries and the very large Girvan grain distillery, as well as having interests in gin, vodka, rum and Irish whiskey. Their Grant's Family Reserve blended whisky is a globally significant brand, selling around 5m cases annually, while Glenfiddich achieves sales around the 1m case mark, making it, by some considerable distance, the world's best-selling single malt. The company is privately held by descendants of the founder, William Grant, and is noted for its independence of outlook.

Loch Lomond operates a significant but low-profile operation at Alexandria, where both grain and single malt distilling is carried out, largely in support of own-label and value brands, such as the company's own idiosyncratically-named Loch Lomond Single Highland Blend. Loch Lomond was in dispute with the SWA, which they have not joined, over the introduction of the Scotch Whisky Regulation 2009 and maintained, unsuccessfully, that a single malt produced from a 100% malted barley wash could legitimately be produced in a continuous still entirely constructed of copper.

The company also owns the Glen Scotia distillery, a small single malt plant in Campbeltown, which produces on an occasional basis. Loch Lomond is privately owned by Sandy Bulloch and family, though it was reported in February 2013 that negotiations to sell the company were at an advanced stage. A significant volume of the value brand, Glen's Vodka, is also produced by Loch Lomond.

Other producers not previously mentioned include J & G Grant, the family owners of Glenfarclas distillery, Angus Dundee, another privately-held producer for the own label market, BenRiach Distillery Company (BenRiach, GlenDronach and Glenglassaugh), Campari, owners of Glen Grant, Beam Inc. (Ardmore and Laphroaig), Gordon & MacPhail (Benromach and substantial third party bottling interests), Isle of Arran Distillers, J & A Mitchell (Springbank and Glengyle), Nikka (Ben Nevis), Morrison Bowmore Distillers (Bowmore, Glen Garioch and Auchentoshan, ultimately owned by Suntory of Japan), Speyside Distillers, Tomatin (owners Takara Shuzo, Japan) and Tullarbardine (owners, Picard Vins et Spiritueux, France) – from which it may be gathered that ownership of the Scotch whisky industry, outside of the major players, is both diversified and international.

In addition to the above, there are a growing number of boutique or craft distilleries, and this sector may be anticipated to grow over the next decade with further entrants. See also chapter 5 for further commentary on this phenomenon.

A number of commentators have begun to refer to the present period as a 'golden age' for Scotch whisky, reminiscent of the boom periods of the 1890s and the 1960s and 1970s. Such an analysis has been criticised as complacent and tending to ignore the impact and greater global success of vodka. In 2009, long-standing critic of the industry, Donald Blair, suggested that *"analysis reveals a true rate of growth that, far from outperforming other global economic and demographic indicators, has seriously lagged the industry's true potential for growth over the last 25 years"*.[27]

This, however potent an analysis (and Blair's work merits rather more detailed consideration than in the opinion of this author it has tended to receive to date), is generally regarded as an eccentric, minority view. The prevailing opinion, energetically expressed by the SWA, is that the performance of the Scotch whisky industry is nothing short of exceptionally strong.

Economic research[28] undertaken for the SWA has concluded that *"today, the total economic impact of Scotch whisky on Scotland's economy is nearly £4.2 billion supporting around 36 000 jobs in the industry and supply chain. Productivity has accelerated to previously unseen levels, this year (2012) the GVA[§§§] per worker was conservatively estimated to be £275,000"*.

"The average Scotch whisky worker adds nearly five times the value added by workers in Scotland's life sciences sector. Workers in the Scotch whisky industry are 57% more productive than workers in London's financial and business services industry".

The report points to exceptional levels of recent investment in Scotch whisky, including the construction of very large new distilleries, the expansion of existing facilities and the re-opening of mothballed plants. Diageo, Chivas Brothers and others have announced very significant investment plans – in Diageo's case amounting to over £1bn over five years, though this figure includes the working capital necessary to support increased stocks, as well as capital expenditure on new distilleries and warehousing. This has been predicated upon sustained increases in prices, both for new fillings and for mature bottled whiskies and by the industry's ability to successfully penetrate new developing markets (see chapter 7).

There is currently little or no evidence of worldwide demand for Scotch whisky slowing down or reversing. The newly-affluent and aspiring middle class consumers of the BRIC nations (Brazil, Russia, India and China), not to mention Africa or the rest of Latin America, appear to have developed a great affinity for premium imported products with strong brand values and an authentic provenance. The demographic trends in these countries strongly suggest continued growth in demand and Scotch whisky is well placed to exploit this for the foreseeable future.

Scotch's market penetration (or 'share of throat', to use the marketing jargon) is still relatively modest in many developed markets: if this could be increased to the levels seen in more mature markets then many of the predictions for future growth may even turn out to be modest.

The whisky industry of the 1890s celebrated 'a big boom', only to founder in dramatic fashion. Over a century later, we may quietly

---

[§§§]GVA: Gross value added; a measure of comparative economic performance.

hope that Scotch's 21st century big boom is well founded and the industry can look confidently to a future of sustained prosperity.

REFERENCES

1. C. MacLean, *Scotch Whisky – A Liquid History*, Cassell, London, 2003.
2. E. Grant, *Memoirs of a Highland Lady*, John Murray, London, 1898.
3. J. R. Hume and M. S. Moss, *The Making of Scotch Whisky*, Canongate, Edinburgh, 2000.
4. Anon, possibly J. Reid, *An Apology for Whisky*, Sands Donaldson Murray & Cochran, Edinburgh, 1759.
5. Anon, *Impartial Observations on the Mode of Levying the Distillery Duties*, Edinburgh, 1786.
6. See *The History of Kennet Pans*, 2011, available online at http://www.kennetpans.info (accessed 29th July 2013).
7. The Farmers of Scotland, *An Address to the Landed Interest of Great Britain on the Present State of the Distillery*, Edinburgh, 1786.
8. D. J. Brown, *Politicians, the Revenue Men and the Scots Distillers, 1780–1800*, in *Review of Scottish Culture 12 (1999–2000)*, Tuckwell Press *et al.*, Edinburgh, 1999.
9. Anon., *Cheap Whisky, a Familiar Epistle to Mr. Pitt on the Recommencement of Distilling in Scotland. To which is Added, The Gowd O' Gowrie: A Scots Song Never Before Published*, Brash & Reid, Glasgow, 1796.
10. G. Skene Keith, *State of Facts, relative to the Scotch Distillery*, A. Brown, Aberdeen, 1798.
11. S. Kelly, *Scott-Land: the Man who Invented a Nation*, Birlinn, Edinburgh, 2011.
12. See, The Scotsman, *More to exporting than bagpipes and tartan*, 2008, available online at http://business.scotsman.com/business/More-to-exporting-than-bagpipes.3652292.jp (accessed 29th July 2013).
13. L. Moyes, *A Letter to the Earl of Liverpool*, J. Booth, Aberdeen, 1821.
14. M. Given, *The Archaeology of the Colonized*, Routledge, London, 2004.

15. A. Waugh, *Merchants of Wine*, Cassell and Company, *London*, 1957.
16. E. Burns, *Bad Whisky*, Angel's Share/Neil Wilson Publishing, *Glasgow*, 2009.
17. Quoted in D. Cooper, *A Taste of Scotch*, Andre Deutsch, London, 1989.
18. A. Barnard, *A Visit to Pattison*, Elder & Co's, Leith and Glenfarclas-Glenlivet Distillery, Sir Joseph Causton & Sons, London, c. 1891.
19. DCL Gazette, April 1924.
20. Anon., to the Editor of *The Daily Telegraph*, 21st January 1903.
21. R. B. Weir, *History of the Malt Distiller's Association of Scotland*, unpublished manuscript, 1974.
22. G. M. Thomson, *Caledonia, or The Future of the Scots*, Kegan Paul Trench Trubner, London, 1927.
23. Speirs and Jeffrey, *A Survey of the Whisky Industry & Quoted Companies*, privately published, Glasgow, 1969.
24. A. S. Gray, *The Scotch Whisky Industry Review*, Sutherlands Edinburgh, Edinburgh, published annually.
25. S. R. H. Jones, *Brand Building and Structural Change in the Scotch Whisky Industry*, Dundee Discussion Papers in Economics, No. 133, University of Dundee, 2002.
26. J. Espey, *A Multinational Approach to New Product Development*, in *Eur. J. Marketing*, 1985, **19** (3), 7.
27. D. Blair, *The Scotch Whisky Industry – Hit or Myth?*, Polestar Scotland, Edinburgh, 2009.
28. 4 Consulting, *Scotch Whisky & Scotland's Economy A 100 Year Old Blend*, Scotch Whicky Association, Edinburgh, 2012.

CHAPTER 3

Crop-to-Cask—Production of New Make Spirit

3.1 INTRODUCTION

From the previous chapter, you will have probably formed the view that whisky is not mass-produced fire-water but, rather, a product carefully crafted over a number of years to achieve the correct balance of visual and flavour qualities. Knowledge of the brand and/or the distillery from which the brand hails will also allow the familiar consumer to establish expectations about the product even before placing an order at the bar or purchasing in the off-trade. We will come back to perceptions of quality in Chapter 5, but it is worth dissecting whisky to understand what makes up its composition.

In all but the highest strength whiskies, water is the most abundant component in terms of percentage volume. Indeed, from a molar perspective, water is invariably the most abundant component in matured spirits (Table 3.1). The provenance of this water may be varied, most or all usually coming from either the streams (or, in Scotland, burns) or mains water nearby the distillery. If maturation, blending and packaging are carried out at a location remote from the initial distillery, then some of the water in the final product will come from a different local supply. Thus, most if not

The Science and Commerce of Whisky
By Ian Buxton and Paul S. Hughes
© Buxton & Hughes, 2014
Published by the Royal Society of Chemistry, www.rsc.org

Table 3.1 Typical proximate analyses of whiskies.[a]

Product	Water (M/L)	Ethanol (M/L)	Water (%(v/v))	Ethanol (%(v/v))	Higher alcohols/esters (%(v/v))
20% ABV "whisky and water"	44	4.3	80	20	0.07
30% ABV "whisky and water"	39	6.5	70	30	0.10
40% ABV grain	33	8.7	60	40	0.04
40% ABV malt	33	8.7	60	40	0.13
55% cask-strength malt	25	12	45	55	0.18

[a]Whilst the presence of components such as wood extractives and sulphur compounds has a substantial impact on whisky flavour, on a weight basis they make a minimal contribution in terms of mass.

all water in whisky originates from the mashing of malted barley, or from the cooking of cereals prior to starch breakdown. Most of this water will be subsequently removed during distillation, with additional losses of water occurring during maturation.

After water, ethanol is clearly the next most abundant component of whisky (Table 3.1). A 40% ABV whisky is around 9 M with respect to ethanol, a rather concentrated solution! But a typical cask-strength whisky will typically weigh in at a mighty 12 M. These values are far in excess of the sensory thresholds for ethanol (thought to be around 14 g L^{-1} or 0.3 M) and of course ethanol is clearly perceptible in whiskies, even when diluted with water or cooled and diluted with ice. This ethanol is derived exclusively from the fermentation of simple sugars, where, for instance, 1 kg of glucose can theoretically yield 511 g (648 mL) of ethanol.

The properties of whisky are then dominated by the ethanol/water matrix. Indeed, it is important to remember that, when mixed, ethanol and water do not behave in an ideal manner. That is, the properties of such mixtures can deviate substantially from a weighted average of such properties. One graphic example is the impact that mixing ethanol and water has on the resulting viscosity (Figure 3.1). From a sensory perspective, this disproportionate increase in viscosity is no doubt perceived as contributing to mouthfeel. It is not clear why such mixtures behave in such a non-ideal way, but it is probably due to a combination of the way in

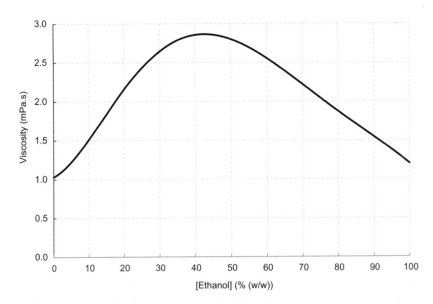

Figure 3.1 The relationship between the composition of ethanol/water mixtures and viscosity at 20 °C. The behaviour is clearly non-ideal, with
peak viscosity reaching levels almost three times that of pure water.

which hydrophobic moieties (such as the ethyl group of ethanol)
confer order on the surrounding water matrix and the observed
ethanol clustering at ethanol concentrations >17% ABV.[1] However, as yet there is no completely satisfactory explanation as to the
cause of this phenomenon. What is clear though from the work of
Macatelli *et al.*[2] is that non-volatile components extracted from
oak casks during whisky maturation influence the structure of
ethanol–water in whisky, which in turn affects the flavour activity
of other volatile flavour compounds in the final spirit.

These other volatile components are principally higher alcohols
and esters. Most if not all of these are by-products of fermentation
and, whilst they do not occur in such high quantities as ethanol,
on a molar basis they are more potent in terms of their sensory
properties and can be well-expressed in final products. Like ethanol, these compounds are formed principally during fermentation
and are subsequently transferred into new make spirit during distillation. The production and fractionation of esters and higher
alcohols into new make spirit and, ultimately, into the final whisky
will be discussed below, but it is worth reviewing these key

Table 3.2 Typical levels of congeners in Scotch whiskies. Compositional data is based on that of Ref. 3, while thresholds are from Refs. 4 and 5. Congeners are quoted in g/100 L of absolute alcohol.

Congener	Whisky		Estimated thresholds (mg L^{-1})	Flavour units[a]	
	Speyside	Islay		Speyside	Islay
Acetaldehyde	5.4	7.0	25	0.43	0.56
Ethyl acetate	26.3	33.2	30	1.7	2.2
Diacetyl	2.1	2.8	0.1	42	56
Methanol	4.8	9.0	10 000	<0.01	<0.01
Propan-1-ol	40.3	37.6	1000	0.08	0.08
2-Methylpropan-1-ol	80.1	85.6	160	1.0	1.1
2-Methylbutan-1-ol	44.2	53.1	65	1.4	1.6
3-Methylbutan-1-ol	139	171	70	4.0	4.9
Total higher alcohols	*304*	*347*	–	–	–
Ethyl lactate	3.7	5.3	250	0.03	0.04
Ethyl octanoate	1.8	2.7	0.4	9.0	13
Furfural	3.6	4.8	150	0.05	0.07
Ethyl decanoate	6.0	8.9	1	12	18
2-Phenylethanol	7.2	8.7	125	0.12	0.14
Ethyl myristate	0.9	1.2	2.5	0.36	0.64
Ethyl palmitate	2.8	3.3	3	1.8	2.2
Ethyl palmitoleate	1.6	2.1	3	1.1	1.4
Phenols	0.01	0.11	[b]	–	–

[a]Calculated on the basis of the dilution of whiskies to 20% ABV. Note that flavour thresholds are highly variable depending on both the matrix of the flavour and the individuals.
[b]Highly dependent on the profile of phenols present.

components found in the final product (Table 3.2). Perhaps the most important aspect to remember is the relationship between the typical concentrations of the various components and their flavour intensity. Such relationships are not exact, not least because of the difficulty in establishing flavour thresholds and their sensitivity to changes in the whisky matrix. Nonetheless, flavour thresholds are a useful guide to what might be expected from the final product in terms of taste.

In common with many foods and beverages, sulphur compounds can substantially affect the flavour characteristics of whisky. These sulphur compounds are derived both from the raw materials and from fermentation. Dimethyl sulphide (DMS) is a volatile compound derived from the degradation of *S*-methylmethionine from malted barley. It is lost from maturing spirit both by evaporation from the oak cask, and oxidation to the relatively flavour-inactive dimethyl sulphoxide (DMSO) and dimethyl sulphone. Both dimethyl di- and tri-sulphides are also found in whisky, the latter, in particular, being an important flavour attribute of some spirits.

The spirit that is filled into casks is water-white, and the entire colour that whisky has at bottle filling is derived from the oak casks in which the component spirits have been matured. This colour comes exclusively from the cask. However, with the exception of water, caramel can be added to whisky to manage colour performance. More specifically, the caramel must be type I (also known as spirit caramel or E150a) for use in whisky as types II, III and IV are not soluble in 70% ABV solutions. The sole reason for adding caramel is to adjust the colour of the final product, which is a reflection of the variability inherent in colour development from oak casks. The typical levels of addition are not generally considered to impact on the flavour characteristics in any way. As we will discuss in Chapter 5, whiskies not bound by the 2009 Scotch Whisky Regulations can contain other ingredients.

It is essential that the alcohol present in whiskies is generated from potable (*i.e.*, agricultural) sources. There are additional restrictions on the carbohydrate sources used depending on the definition of whisky in a given part of the world. The use of 'chemically synthesised' ethanol, for instance by addition of water to ethene (in turn derived from oil cracking) is not permitted. Whilst the alcohol in alcoholic beverages must be derived from agricultural sources, plant substrate is invariably carbohydrate in origin. Such carbohydrate can be considered to be either fermentable or non-fermentable. In the first case direct extraction of sugars and subsequent fermentation suffices, but where the carbohydrate is not fermentable it must be converted into an extractable derivative, which affects the design and operation of the manufacturing process.

Fermentable sources, mainly derived from fruits, sugar beet and sugar cane, can be fermented directly, although the use of such substrates for the production of whiskies is generally limited to molasses, which is primarily a by-product of sugar cane processing. Molasses is a common substrate for many alcoholic beverages, including some rums and Indian whisky.

The most common raw materials that do require some form of carbohydrate hydrolysis prior to fermentation include malted and unmalted barley and wheat, maize, oats, cassava, sorghum and agave. The carbohydrate source of the latter is inulin, a biopolymer of fructose, and its use, wholly or in part, is restricted to Tequila production. Cassava is problematic both in that it can be appreciably cyanogenic, releasing hydrogen cyanide during processing, and it can yield too high levels of isopropyl alcohol, a higher alcohol that is difficult to remove effectively because of the proximity of its boiling point (80 °C) to that of ethanol (78 °C). Sorghum is also used in some parts of the world, notably India. Here, we will limit the discussion to the processing of starch, primarily from barley, wheat and maize sources, while recognising that other fermentable substrates are also used worldwide.

3.2 WATER AND CEREALS

3.2.1 Water

The raw materials employed should of course be suitable for the production of a potable product. A reliable supply of water is needed for a wide range of duties in whisky production, including malting (when required), mashing and adjustment of alcohol strength. Water is also required for various services, including the raising of steam and cooling. Water for whisky production should be visually clear with no colour or flavour taints and be sufficiently clear of microbial contamination to be considered potable. As we will see below, there is also a requirement for certain minerals, not least calcium. To ensure fitness-for-purpose, particles and chemical contaminants may be removed by filtration and carbon treatment, respectively, whilst micro-organisms can be nullified by treatment with UV light or removed by sterile filtration. In the event that the availability of a suitable water supply is a significant risk for a distillery, it is advisable to have an alternative supply available.

The presence of non-volatile salts has various effects both on the production of whisky and its final quality. Mashing-in water requires minerals, especially calcium ions for α-amylase activity, whilst sulphate and carbonate affect the pH of mashing-in water and trace metals facilitate fermentation. The presence of nitrates is undesirable as they are markers of water contamination. Water has become an increasingly valuable resource and efforts are being made to minimise water usage. Much can be done to tighten up the process to reduce water losses, for instance by automating cleaning-in place (CIP) systems and more careful monitoring of where water losses occur.

3.2.2 Fermentable Carbohydrate Sources

3.2.2.1 Barley. Whilst it is possible to use, in principle, unmalted barley to produce fermentable extracts directly, the technology for converting barley into malt is as venerable as beer making itself. In Scotland, barley malting has evolved along with distilling and indeed, even today, some Scotch whisky producers retain their own maltings (*e.g.*, Highland Park, Balvenie) that produce at least part of their own malt requirements. For over 300 years the importance of selecting the right barley varieties for malting has been recognised. Today, the highest priority characteristics for selecting new malting barleys include:

- Resistance to kernel splitting, a phenomenon that exposes the starchy endosperm to damage
- A mealy (*i.e.*, friable) rather than a steely endosperm, which in turn facilitates rapid and even hydration
- High starch content, with the corollary that there is relatively low nitrogen assimilation in the field
- Vigorous germination for satisfactory malting
- Low glycosidic nitrile potential (see Chapter 5)

Other properties that also impinge on the selection of barleys for malting include homogeneity of kernel size and ripening, and a rather contradictory requirement for both low dormancy and no pre-germination. Today, it is fair to say that the supply chain, from breeders to distillers, works together to develop successful varieties.

A common method to gauge the efficacy of a cereal for distilling purposes, including malted barley, is to determine the predicted spirit yield (PSY; *i.e.*, quantity of pure alcohol) per tonne (dry weight) of the substrate. Over the past 100 years this spirit yield has increased by around 50%, from a relatively modest 300 L tonne^{-1} to more than 450 L tonne^{-1}, although typical minimum specifications for malted barley are of the order of 408 L tonne^{-1}. These improvements are due to improved agronomic practices and breeding of higher yielding varieties. With malted barley costs typically around £450/tonne, this equates to a real reduction in malt costs from £1.50 to a figure approaching £1 per litre of alcohol. The choice of barley varieties for making malt is dependent both on pragmatic economic considerations and the use to which the malt will be put. For malt spirit, the malted barley must have a high PSY. In turn this means that a high hot water extract, low gelatinisation temperatures, rapid and complete modification, and high fermentability of unboiled wort are also of high priority. Other aspects that are of relevance to the maltster include low levels of mycotoxins, low carbon losses during malting and, from a processing perspective, manageable ratios of pentose to pentosans. This latter requirement influences the early phase of barley malting.

Two other factors to bear in mind when selecting barley for malting are the presence of pre-germinated grains and ensuring that the total nitrogen (TN) content is sufficiently low. The former is an undesirable condition, where the grains start to germinate before harvesting, which in turn impacts on the homogeneity of the malted barley. Typically, specifications aim for less than 5% pre-germinated barley in a batch. The desire for low nitrogen levels in grains is understandable in that TN levels negatively correlate to starch content. Several methods are available for the determination of nitrogen, for instance the so-called Kjeldahl and Dumas tests, and in both cases multiplying the nitrogen content by a factor of 6.25 gives an approximate total protein level.

Harvested barley can be pre-cleaned prior to drying, although this is more difficult for wet barley and so maltsters tend to prefer to clean after drying. Barley is dried to around 12% moisture before being stored. The barley is not cooled, so as to facilitate the breaking of dormancy. For many varieties temperatures up to 25 °C are sufficient, but some are less tractable and may need to be stored at closer to 30 °C to ensure that dormancy is broken. The progress of dormancy breaking is monitored by regular assessment of germination energy (*i.e.*, the percentage of grains that can be expected to germinate) and comparing this with the germinative capacity (percentage of living grains). When dormancy is considered to have been broken, the bulk grain is then cooled to around 5 °C for longer term storage. The storage stability of barley is affected by a number of parameters, but especially its moisture content and storage temperature.

Before malting, the cleaned barley is screened on the basis of size. Kernels that pass through a 2.2 mm screen are usually sold for feed, whilst those kernels retained on a 2.2 mm screen are further separated on a 2.5 mm screen. Most of the kernels are retained, but those that pass through, a minor fraction of the whole, can be malted separately as they malt more rapidly than the larger kernels.

3.2.2.2 Barley Steeping. The conversion of barley (Figure 3.2) to malted barley is a critical stage in the development of a fermentable extract from barley. Malting has been extensively reviewed (see Ref. 6) and is summarised in Figure 3.3. The first stage, steeping, is required to initiate the germination of the barley kernel and to hydrate the starchy endosperm sufficiently to enable its modification. Although the germination of barley for malting has obvious similarities to that for plant growth in the field, it is important that carbon is not lost unduly when preparing for malt production as this will adversely affect ethanol yields.

The uptake of water by barley kernels is clearly important as a precursor to germination. Typically, to ensure the even progress of malting, a pattern of wet steeps and air rests are instigated. An initial wet steep elevates moistures to around 32–35% water. Air resting with ventilation allows residual surface water to be taken up. Wet steeps and air rests cycle until the required moisture

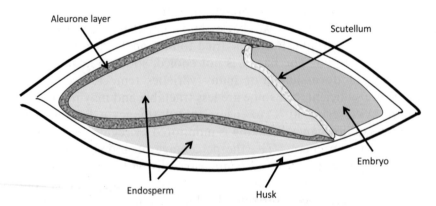

Figure 3.2 Simplified cross-section of a barley grain.

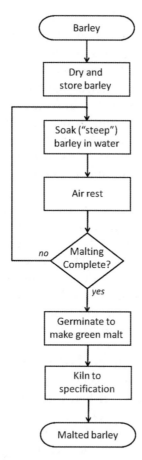

Figure 3.3 Flowchart outlining the main aspects of the barley malting process.

levels are attained. Water uptake occurs in three more-or-less distinct phases, with the first proceeding rapidly up to 32–35% moisture. A second period of water uptake is slow before a third phase of rapid uptake. Overall, steeping time takes a little in excess of two days. It is essential that the endosperm is hydrated adequately before it can undergo modification. All malts contain non-uniformities within the endosperm structure, although some technological approaches to low-grade or poor barleys can offset some of these potential heterogeneities.

The presence of abscisic acid (ABA) in barley acts as an inhibitor to the germination of barley, as it inhibits the development of α-amylase in the aleurone layer. Thus, germination starts once the levels of ABA drop to a low level. During steeping, ABA is washed from the barley into steep water and is sensitive to oxidation, so that it is lost more quickly from aerated steeps. Oxygen is rapidly taken up from the steep water as the barley respires, so there is a concomitant increase in carbon dioxide levels in the steep. Ventilation during wet steeping and air resting helps to replenish oxygen, remove carbon dioxide and, by provision of oxygen, reduces the accumulation of ethanol. Air rests also help to dissipate the undesirable excess heat of aerobic respiration.

3.2.2.3 Barley Germination.   Once the barley grain is properly hydrated it needs to be germinated. The main purposes for the production of distilling malt are to ensure both maximum fermentable extract and amylolytic enzymes. The onset of germination is apparent once the roots and shoot become apparent in the kernel. The shoot grows underneath the husk, clearly visible as a distortion of the husk, and is indicative of enzyme synthesis. Whilst still a subject of active research, it is generally agreed that most enzyme activity is synthesised in, and secreted from, the aleurone layer, modulated in turn by the hormonal gibberellins secreted by the embryo. Gibberellic acid (GA) positively affects the production of α-amylase and limits dextrinase, whilst β-amylase is unaffected by the activity of GA.

Virtually all of the fermentable extract is derived from the starchy endosperm. This extract, derived through a process of modification, is released by a combination of protease and amylase enzymes. The starch, present as both large (*ca* 30 μm diameter) and small (1–5 μm diameter) grains are embedded in a protein matrix within the endosperm cells. These cell walls are composed mainly of glucan and pentosan, which are broken down by enzymes such as β-glucanases and pentosanases that are produced in the aleurone layer. The structural cell wall and protein are significantly degraded during germination, but most of the starch remains intact. Thus, good modification results in an endosperm that has exposed starch, as this ensures that starch will be accessible to enzyme activity during germination.

The ideal conditions for germination are complex. Higher germination temperatures tend to yield malts with lower extract yields, although higher moistures can aid recalcitrant barleys to modify due to higher levels of hydration in the grain. Oxygen is important in the early stages of germination for modification, although carbon dioxide is inhibitory.

There are four key enzyme activities that are required for satisfactory distilling malt: α-amylase, β-amylase, debranching emzymes (*i.e.*, pullulanase or limit dextrinase) and α-glucosidase (or maltase). Whilst they are all active in cleaving glycosidic linkages, they have different modes of action that complement each other to some extent (Table 3.3, Figure 3.4). Thus, whilst the endo-enzyme, α-amylase, cleaves α-1,4 linkages, it is unable to attack α-1,6 linkages, limiting its ability to fully degrade the branched amylopectin. More specifically, complete α-amylase activity will yield dextrins of up to 12 glucose units, with only slow production of fermentable sugars. In contrast, β-amylase is an exo-enzyme and cleaves α-1,4 linkages to release maltose from the non-reducing end of α-1,4-linked chains. Limit dextrinase has utility in that it cleaves α-1,6 linkages and can therefore 'debranch' amylopectin and increase the fermentability of the resulting malt. Maltase liberates β-D-glucose, either from α-1,4-linked chains or terminal glucose residues linked by α-1,6 linkages. It is clear that these complementary activities, when working together, can maximise the fermentability of the resulting malt.

3.2.2.4 Kilning and Peat. The kilning of malt aims to:

- Reduce moisture content to a level that will result in a stable product for storage
- Arrest biological activity

Table 3.3 Major enzymes responsible for the degradation of starch during mashing.

Enzyme	Mode of activity
α-Amylase	Endo-activity cleaving 1-4 links
β-Amylase	Maltose release from non-reducing termini
α-Glucosidase (maltase)	Glucose release from non-reducing ends; 1-4 or 1-6 cleavage
Debranching enzyme (pullulanase or limit dextrinase)	Cleaves 1-6 links to remove branch points
Phosphorylase	Glucose-1-phosphate release from non-reducing termini

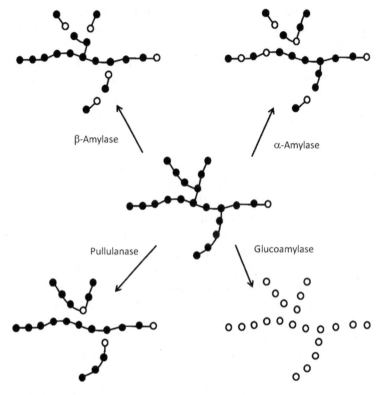

Figure 3.4 Points of attack by the major amylolytic enzymes. Open circles represent reducing termini of the sugar moieties.

- Develop desirable flavours (including those from peat)
- Eliminate unwanted flavours

At the onset of kilning, the evaporation of moisture from the grain bed maintains the malt temperature below that of the warm air applied. This free-drying stage prevails until the moisture content drops to 20–25%. The so-called break is characterised by increasing grain bed temperatures, as the residual water becomes increasingly strongly adsorbed to the grain. By the time moisture levels of 10–12% are achieved, the remaining water is bound within the grain and requires higher drying air temperatures to achieve storage moistures of 4–6%. During the free-drying stage, the grain temperatures remain sufficiently low so that modification and enzyme syntheses continue. However, after the break point, some enzyme activity is lost by denaturation, especially towards the end of kilning.

Because of the necessity for high fermentabilities and PSY, it is essential to try to minimise enzyme activity losses during kilning. In particular it is important to keep initial air drying temperatures to a minimum as enzymes at a given temperature are more labile at higher water activity levels (which prevail at the onset of kilning). This is most acute for the more thermolabile enzymes. With the advent of indirect kilning in Saladin boxes, whereby combustion gases are not passed through the grain bed, fementabilities of laboratory worts have increased significantly from around 86% up to 89%. As well as enzyme activities, kilning affects the levels of other compounds present in the malted barley. Colour development is principally due to the formation of melanoidins from free amino nitrogen and reducing sugars. Such reactions are more rapid when water activities are higher and so will be more pronounced when initial drying is at higher temperatures. Such Maillard reaction processes yield other flavour-active compounds, such as furans, pyrroles and diketones.

Thermal degradation of *S*-methylmethionine yields the flavour-active and volatile DMS. Some of this is carried through into the distilled spirit, where it is lost during maturation (see Chapter 5). The oxidation of unsaturated lipids in malt, particularly linoleic (18 : 2) and linolenic (18 : 3) acids can yield low levels of highly flavour-active aldehydes, such as hexanal (green, grassy) and *trans*-2-nonenal (cardboard).

The presence of sulphur dioxide in kilning is derived either from the sulphur in fuel oils, or introduced directly from gas-fired kiln gases in direct fired kilns. Sulphur dioxide is highly soluble in water and readily taken up by malt, especially during the free drying stage of kilning. The deliberate introduction of SO_2 is to counteract the presence of nitrogen oxides, which are undesirable as they react with compounds such as dimethylamine to form the carcinogenic *N*-nitrosodimethylamine. Such approaches have essentially eliminated the presence of elevated levels of nitrosamines in malted barley today.

For grain distilling, there is the option of using green (*i.e.*, unkilned) malt. This is advantageous in that the costs of kilning are avoided, so that the green malt has a lower cost than its dried counterpart. Additionally, it retains up to 50% more enzyme activity, which is lost when distilling malt is kilned, even at mild temperatures. At least one grain distillery in Scotland continues to use green malt. Of course, given that one of the benefits of kilning is to create a product with appreciable storage stability, it is imperative that the logistics of green malt supply are highly reliable, which in turn means that transport charges are higher.

One of the most iconic raw materials used for whisky production is peat. Peat is derived from the anoxic decay of dead plants. The anoxic condition, often brought about by water-logging, is essential if the plant material is not to be fully decomposed, primarily to CO_2 and water. Peat can be considered to be an intermediate in the formation of first brown, then black coals. As peat development advances, the chemical composition is characterised by loss of relatively labile functional groups, such as amino, carbonyl and hydroxyl residues, and increasing aromatisation of the carbon skeleton. Historically, peat was used as a heating source, but today its role is to provide characteristic peaty flavours (Table 3.4). Whilst not all malt for whisky production is peated, when it is used, it adds to the peat "fire box" when malt is around 50% moisture. Optimum absorption of peat-derived volatiles is at 15–30% moisture. When the grain moisture is below 15%, phenolic adsorption rates decrease. It is important to control the level of oxygen to avoid combustion rather than volatilisation of flavour-active peat-derived compounds and, because of the impact on the flavour of the final product, there is a clear need to manage peating carefully to achieve consistent results. Whilst final levels of peat in malt can

Table 3.4 Common phenolics found in peated malts.

Compound	Sensory attributes
Phenol	Carbolic
2-Methylphenol (*o*-Cresol)	Musty, medicinal
3-Methylphenol (*m*-Cresol)	Woody, ethereal
4-Methylphenol (*p*-Cresol)	Medicinal
Dimethylphenols (Xylenols)	Medicinal, sweet
2-Methoxyphenol (Guaiacol or creosol)	Medicinal, woody, smoky
4-Ethylguaiacol	Smoky, meaty
4-Vinylguaiacol	Spicy, clove
4-(2-propenyl)guaiacol (Eugenol)	Cinnamon, clove, spicy

vary from a few mg kg^{-1}, some of the most heavily peated malts can be as high as 80 mg kg^{-1} of phenolics.

3.2.2.5 Maize. Historically, maize was the main raw material for the production of grain whisky in Scotland, not least because it was a cost-effective and relatively easily-accessible source of starch. However, trade barriers, such as increases in import duties into the EU, made wheat a more cost-effective option, although today maize does come into Scotland from the south of France. Whilst maize has, in the past, been purchased based on the logical specifications of specific weight and moisture content, the huge increase in the proportions of genetically-modified (GM) maize crops now grown dictates that distillers have to select supplies that are certified as GM-free.

Maize is less problematic to process than wheat. It gives a higher spirit yield (typically maize is 71–72% starch and 10% protein), and contains less of the highly viscous pentosans and glucans. Maize can be used with or without the embryo or germ. The starch in maize is more refractory and requires higher processing temperatures to ensure adequate gelatinisation of the starch. As is the case for barley, the starch in maize is embedded in a protein matrix within a cell wall structure. At the core of the maize kernel is a soft, floury starch, and this is surrounded by a harder, glassy starch. It is this harder, glassy structure that dictates the need for processing at high temperatures and pressures. There is an ongoing debate concerning the flavour attributes of grain spirit produced using wheat or maize, but suffice to say that whilst the resulting spirits are ostensibly similar analytically, they do seem to show different flavour performances, with wheat perhaps contributing to a lighter-bodied spirit.

3.2.2.6 Wheat. Today, wheat is a common substrate for the production of grain whisky and requires quite different processing to maize for the production of grain spirit. One positive aspect of using wheat is the *plethora* of varieties that are available. However, from the perspective of alcohol yield, soft red or white winter wheats are the best-performing. They also give less processing problems as they give worts of lower viscosity than those from hard wheats, which in turn affects process-limiting steps, such as wort transfers and the evaporation of spent wash. Distillers can purchase soft wheat essentially at commodity prices as they are not in competition with bakers, who require hard varieties with high nitrogen. Riband has been popular with distillers because of its excellent and consistent distilling yield performance and, indeed, despite being outstripped by new wheats, it retains its popularity. Other varieties, such as Denman, are likely to herald the future.

3.2.2.7 Barley. The use of unmalted barley has obvious attractions as it is inexpensive in comparison with malted barley, wheat and maize. However this belies challenges in its direct use, not least of which are the presence of high levels of gums and β-glucans that can result in severe processing problems. Unlike beer production, Scotch whisky cannot be produced using exogenous enzymes, and so at least in Scotland the use of raw barley is essentially precluded under the current Regulations.

3.2.3 Preparing the Fermentable Extract

As has already been mentioned, the ultimate source of fermentable extract is carbohydrate-derived and, for Scotch whisky, the main sources are malted barley, wheat and maize (whole or degermed). These sources are, though, essentially solid materials and hold the sugars that yeast will require in the form of starch. Yeast cannot act directly on starch, so to release fermentable carbohydrate, the solid must be physically broken down and the starch degraded. To liberate fermentable sugars from starch, three processes are required:

1. Disruption of the endosperm cell walls
2. Explosion of the tightly-packed starch granules to expose starch to amylolytic enzymes
3. Enzymatic degradation of the exposed starch to generate a fermentable sugar extract.

In one sense, this is akin to aspects of the germination of the grain that develops into a fully-grown plant. However, from an industrial perspective, gelatinization is essential to accelerate the exposure of starch to enzymic activity. The gelatinization of starch is brought about by a combination of temperature and pressure.

Table 3.5 Properties of starch sources commonly used in whisky production.

Starch	Starch granule size/μm	Typical gelatinisation temperature/$^\circ C$
Malted barley		
– Small starch granules	1–5	61–62
– Large starch granules	10–25	75–80
Maize	10–15	70–80
Wheat	1–5, 15–25	52–54

The exact conditions required will depend on the starch source (Table 3.5) and indeed, for malted barley, the conditions required for gelatinization are a function of the size of the starch granules themselves as small starch granules tend to be more refractive than large granules. It is essential that gelatinized starch is comprehensively degraded by the enzymes present. The consequences are two-fold: firstly, gelatinized starch, when cooled, forms retrograded starch, a tough material that is resistant to enzyme degradation and is difficult to re-gelatinise. Secondly, the presence of retrograded starch implies a loss of alcohol yield, thus reducing the efficiency of production and therefore increasing costs per litre of alcohol produced.

The distiller is naturally keen to produce alcohol consistently and efficiently and is therefore focused on creating a consistent fermentable extract. However, whilst yeast can convert fermentable sugars into alcohol, the yeast itself also needs to grow, requiring in turn a source of nitrogen for protein synthesis and lipids for membrane synthesis. The nitrogen source is principally amino acids derived from the proteolysis of cereal-bound proteins. For Scotch whisky production, all nitrogen must be derived from the raw materials, rather than from the addition of exogenous sources. In common with the amylases, the key proteases in malted barley also operate optimally under a range of conditions (Table 3.6). Free amino nitrogen also serves another purpose, providing a source of major congeners in distilled spirits (see below).

3.2.4 Fermentable Malt Extracts

To expose the starch and proteins to hydrolysing enzymes, the raw materials are typically broken up by milling. The exact conditions of milling are dependent on both the requirement, or not, to retain

Table 3.6   Summary of malt-borne proteases and
               their pH optima.

Enzyme	pH optimum
Endoproteases	
Thiol-dependent enzyme 1	3.9
Thiol-dependent enzyme 2	5.5
Metalloenzyme 1	5.5
Metalloenzyme 2	6.9
Metalloenzyme 3	8.5
Exoproteases	
Five carboxypeptidases	4.8–5.6
Four neutral aminopeptidases	*ca* 7.2
Two alkaline peptidases	8–10

an intact husk for subsequent removal, as is often the case for malt
spirit production, and indeed the raw material itself. Before milling
though, it is essential to clean the grain, especially to ensure removal of stones and metal fragments, which can both damage the
mill rollers and create a spark or even explosion hazards. Ferrous
metals are typically removed by magnetic separation, whilst stones,
generally being of higher density than grains, can be effectively
removed by gravity. Other contaminants, such as seeds, can also
be removed depending on the cleaning system installed. Malted
barley is typically milled using a four-roller mill. The first two pairs
of rollers effectively squeeze the kernels between the first pair of
rollers and the second pair fragments the endosperm further.
Milling can be performed with one pair of rollers but the additional
two rollers are important as a factor for maximising extract yield.

The milled malt is generally prepared as required as it can readily
absorb moisture from the air. It is mixed with hot water so that the
overall temperature is 'optimal' for extract production. The term
optimal should be interpreted with caution because, as we have
discussed above, the range of enzymes that are required for effective conversion requires a compromised set of conditions, especially in terms of temperature and pH. In practice the mashing-in
temperature is often around 63–64 °C, with a pH of around 5.2.
Temperature control is critical here as relatively small changes in
temperature can adversely affect the extract yield and, therefore,
the subsequent alcohol yield. After the mashing stage is complete,
the insoluble material from the malted barley, known as draff, may

or may not be removed before it is cooled and transferred to the fermenter.

3.2.5 Fermentable Grain Extracts

The production of a fermentable grain extract tends to be more variable than the equivalent process for malted barley. Generally, the enzyme complement is provided by a mere 10–15% of the total grist, and so the onus is on the selection of malt with sufficiently high enzyme content to effectively convert the starch from all of the raw materials into fermentable sugars.

Although there are relatively few grain distilleries, each has its own proprietary process for production. Generally, the grist needs to be prepared by milling, followed by cooking of the grist, discharging the cooker and finally conversion of the cook to a fermentable extract. Milling is dependent on the grist, so that roller or hammer mills are commonly employed, although, in the case of green malt, wet milling can be a useful alternative. Hammer milling does have the advantage of providing a grist that gives higher extract efficiencies and faster cooking times, although a fine grind results in particles with relatively high surface free energy, which in turn gives rise to particle agglomeration or 'balling'. An excess of fines can also add additional burden on downstream processing of spent wash. Some distillers choose not to mill the cereal feed prior to cooking. Efficiencies of processing in terms of milling losses, however, may be more than offset by losses of yields from subsequent cooking.

As mentioned previously, wheat is the predominant feed for grain distilleries in Scotland. Although wheat gelatinises at relatively modest temperatures, it seems that higher extract yields can be derived from the cooking of wheat. It is also fair to say that grain distilleries are likely to install cooking facilities when required to switch between maize and wheat.

Cooking at elevated temperatures and pressures allows for the efficient gelatinization of starch and indeed when the cooked slurry is removed from the cooker, the reduction in temperature and pressure, known as blow-down, can effectively pop the residual grain. It is important, however, to ensure that the starch does not crystallize as retrograded starch as this is resistant to enzyme hydrolysis and therefore represents a loss in extract and potential processing problems. The presence of higher levels of the more structurally-regular amylose increases the risk of retrogradation and, unfortunately, wheat and maize starches can typically contain around 26% amylose. By blowing down rapidly to around 70 °C and then adding the malt, the opportunity for retrogradation is minimised. Mashing can then proceed in the usual way, with the amylases and proteases degrading their respective biopolymers to create a fermentable extract.

3.3 FERMENTATION

The key role of fermentation is to produce ethanol and congeners that will ultimately flavour the new make spirit and be expressed as flavours in the final whiskies. No matter what the source of the raw materials or the mode of distillation, fermentation contributes breakdown products and metabolites of yeast and bacteria. Their contribution may be more subtle in more highly rectified spirits, such as those from continuous distillation, but nonetheless the fermentation can, with volatiles from the raw materials and extractives from wood, be considered to be the major flavour contributors to all whiskies. Indeed, this drives the stipulation for Scotch whisky that the ethanol content of new make spirit cannot exceed 94.8% ABV so that at least some of the features of flavour from the raw materials and fermentation are retained.

Historically, distillers have relied on the excess yeast produced by brewers, but developments in the breeding of yeast hybrids allowed distillers to bring together the classical properties of the ale yeast, *Saccharomyces cerevisiae*, with *Saccharomyces diastaticus*. The latter was discovered as a brewing yeast contaminant, with an ability to ferment beyond the trisaccharides, for instance, maltotetraose and some of the otherwise unfermantble disaccharides. The Scotch Whisky Regulations state that the breakdown of cereal starch must be affected only by enzymatic activity of malt, but the additional fermentability conferred by the *S. diastaticus* hybrid is accepted. As well as having desirable and flavour-producing characteristics, distilling yeast must also perform adequately in the process. Specifically, distilling fermentations are performed at elevated temperatures to achieve rapid completion even in the presence of high concentrations of ethanol, so distilling yeast should exhibit growth above 30 °C and function up to 35 °C. Additionally, as the final wash should be 7–9% ABV, distilling yeast should be osmotolerant, withstanding wort strengths of 16–20 degrees Plato. A consequence of the need for rapid and complete fermentation is that as much yeast as possible should remain suspended in the fermenting wort, and so it is essential that distilling yeast is non-flocculent. Finally, wort fermentations are prone to foaming and the higher the fermentation temperature the greater the foaming, which in turn occupies fermentation capacity. As whisky production generally precludes the use of antifoams, distilling yeast ideally should not be prone to foaming in the fermenter, although rotating foam breakers, or so-called 'switchers', at the top of the washback can effectively break up drying foam.

Whilst distillers demand a consistent and high-quality spirit, they also want as high an ethanol yield as possible from their fermentations, which implies avoiding excessive yeast growth. Distilling-yeast sources its nitrogen requirement from the various amino acids and small peptides released during cereal malting and mashing. Yeasts also require other elements, such as phosphorus and sulphur (available from wort in the form of phosphates and sulphate), as well as a *plethora* of metal ions, most importantly, potassium, zinc, manganese and iron. Yeasts are able to synthesise all their required vitamins, with the exception of biotin, although malt and grain worts contain sufficient levels of biotin to ensure reliable fermentations. Ethanol is only produced during anaerobic respiration; however, it is essential, practically, to aerate wort for the purpose of fatty acid synthesis as wort does not contain sufficient fatty acids to satisfy all of the requirements for yeast growth and membrane synthesis.

Ethanol and carbon dioxide are the main products of fermentation by mass, but nonetheless around 400 other compounds are produced during alcoholic drink fermentations. As discussed earlier, the quantity of chemical components by mass is less important than how it relates to the flavour threshold. The encyclopedic work compiled by Nykanen and Suomalainen[7] lists the major products from fermentation (Table 3.7).

3.3.1 Key Metabolic Aspects of Distilling Fermentations

The carbon of ethanol produced during fermentation is derived from fermentable sugars. These sugars must be taken up by the

Table 3.7 Major products of yeast in distilling fermentations. (After Refs. 7 and 8.

Alcohols	Acids	Esters	Sulphur compounds	Others
Ethanol	Acetic	Ethyl acetate	Hydrogen sulphide	Carbon dioxide
Propan-1-ol	Hexanoic	3-Methylbutanoyl acetate ("isoamyl acetate")	Dimethyl sulphide	Acetaldehyde
Butan-1-ol	Octanoic	… and others	Dimethyl disulphide	Diacetyl
3-Methylbutan-1-ol ("isoamyl alcohol")	Lactic		Dimethyl trisulphide	
Glycerol	Pyruvic			
β-Phenylethanol	Succinic			

Table 3.8 Sugar uptake and permease activity in a typical 48 h distilling fermentation.

Time/h	Permease activity	Sugar uptake
0	↑ Glucose permease	
6		Glucose, fructose, sucrose
12	Maltose permease	Glucose, maltose, fructose, sucrose
18		Maltose
24		Maltose
30	↑ Maltotriose	Maltose, maltotriose
36	permease	Maltotriose
42		Maltotriose
48		Maltotriose

yeast cell, converted into ethanol, carbon dioxide and flavour congeners within the cell, and the products excreted into the wash and potentially recovered in the new make spirit and the final whisky. Malt and grain fermentations have similar sugar profiles, with maltose being the dominant sugar, along with sucrose, fructose and glucose and higher order sugars. Their mechanism of uptake is also specific to the identity of the sugar (Table 3.8).

The metabolic pathways for the production of ethanol (Figure 3.5) are well described in many biochemistry texts, and by Ref. 9. Briefly, for the Embden–Meyerhof–Parnas (EMP) pathway,

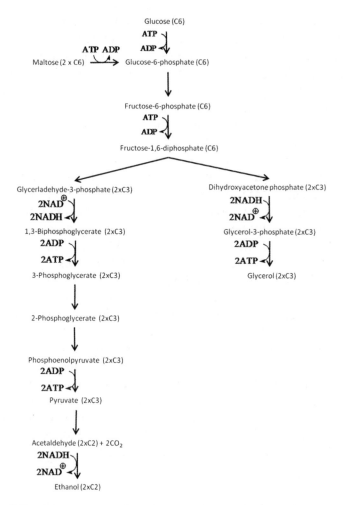

Figure 3.5 An overview of the Embden–Meyerhof–Parnas pathway that represents the major route to ethanol synthesis by distilling yeasts.

glucose and fructose are phosphorylated, whilst maltose is diphosphorylated during uptake before hydrolysis. In any case, glucose-6-phosphate is isomerised to fructose-6-phosphate. This is essential as the latter can be split to yield two identical molecules of glyceraldehyde 3-phosphate, which is converted to pyruvate recovering additional molecules of adenosine triphosphate (ATP). Pyruvate undergoes decarboxylation to yield carbon dioxide and acetaldehyde, which in turn is reduced to ethanol by NADH.

An alternative pathway, exemplified by homofermentative lactic acid bacteria, sees pyruvate reduction to lactate without the loss of carbon as carbon dioxide. A minor alternative pathway terminates in the formation of non-volatile glycerol. In terms of flavour congeners, the EMP pathway yields mainly ethanol and acetaldehyde, but other compounds, such as higher alcohols and esters, are produced as by-products from other yeast biosynthetic pathways. Whilst this discussion has focused on the EMP pathway, it is important to appreciate that other pathways can exist, especially under aerobic conditions at the beginning of fermentation.

The presence of free amino nitrogen (FAN) is essential for both yeast growth and for flavour development. The amount of FAN present is dependent both on mashing-in conditions and the degree of modification of the malt. Excessive FAN levels stimulate yeast growth to an extent that there are appreciable losses of carbon into cell growth rather than ethanol production. Yeast cells do not have active uptake mechanisms for all amino acids, and the venerable Jones and Pierce[10] classification persists today as a useful guide to the uptake rates of amino acids by yeast (Table 3.9). Aspartate and asparagine, glutamate and glutamine, and lysine do have dedicated transport enzymes. Arginine, serine and threonine share a single permease, but all of these aforementioned amino acids, group A in the Jones and Pierce classification,[10] are taken up at a sufficiently rapid rate to satisfy the requirements of the growing cell. The B group share a common permease and are slowly absorbed. Group C, relying on a general permease for transport, are only taken up

Table 3.9 Classical Jones and Pierce classification of amino acid uptake from brewers' wort.[10]

Group A	Group B	Group C	Group D
Rapid absorption at onset of fermentation	*Slow absorption from start of fermentation*	*Slow absorption late in fermentation*	*Slow absorption during fermentation*
Aspartate/asparagine	Histidine	Alanine	Proline
Glutamate/glutamine	Isoleucine	Glycine	Hydroxyproline
Lysine	Leucine	Phenylalanine	
Arginine	Methionine	Tryptophan	
Serine	Valine	Tyrosine	
Threonine			

late in fermentation, too late for significant cell growth. The final group D are absorbed slowly, but at rates sufficient to satisfy the biosynthetic needs of the yeast cell. As there is a general deficiency of groups B and C, the yeast cell must supplement these amino acids *via* additional biosynthetic pathways. In fact, lysine cannot be utilised for this purpose, so the yeast cell relies on glutamate and glutamine, and aspartate and asparagine to fulfil this role. However, if the supply of amino nitrogen is depleted, the intermediate α-keto acids are readily decarboxylated to their corresponding aldehydes and then reduced to the corresponding higher alcohols (a mechanism analogous to the conversion of pyruvate to acetaldehyde and ethanol; Figure 3.6).

One of the most important flavour congeners formed during fermentation is diacetyl. This is formed by the chemical oxidation of secreted acetoin and 2,3-butanediol, which in turn are derived from acetolactate, an intermediate in the synthesis of group B amino acid, valine (Figure 3.7). Whilst the conversion rate to diacetyl is low yielding, it is of disproportionate importance both because of its low flavour threshold in spirit and the fact that its

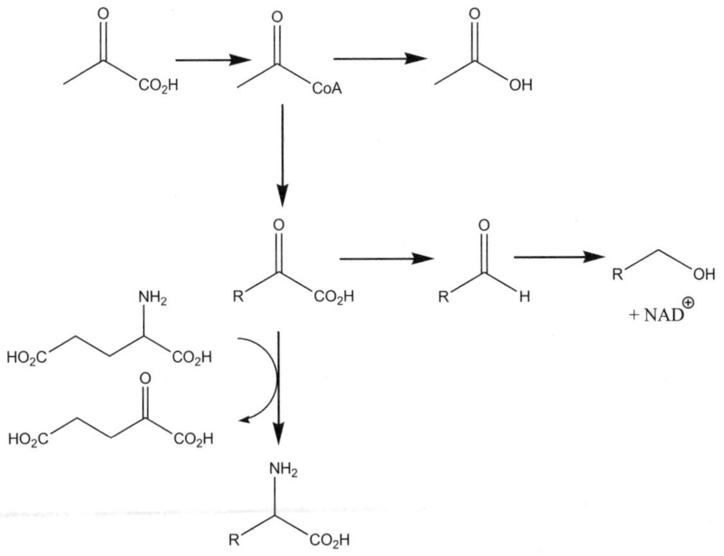

Figure 3.6 Degradation of excess α-ketoacids to yield aldehydic precursors of higher alcohols.

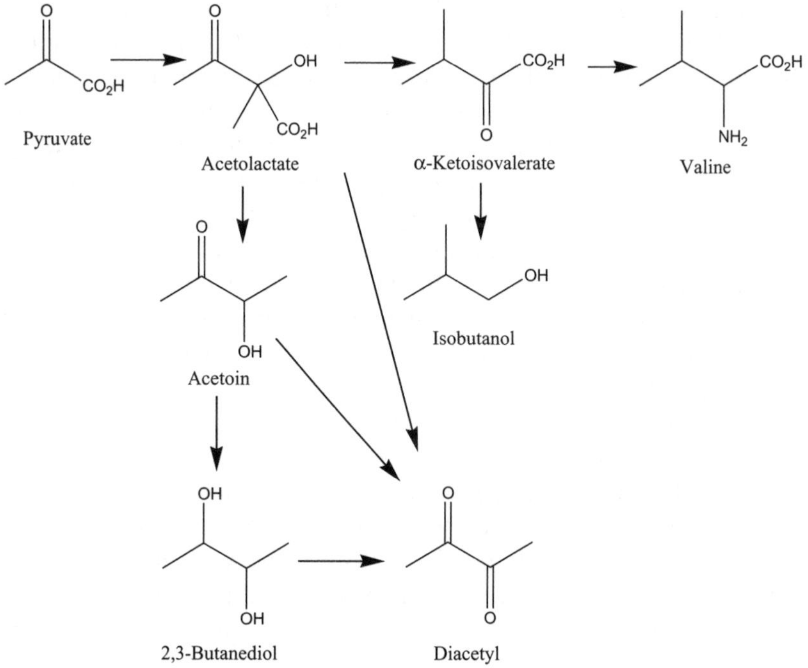

Figure 3.7   Formation of diacetyl as a by-product of valine biosynthesis.

volatility is so similar to ethanol that it is not possible to distil diacetyl from ethanol completely.

Esters are also very important flavour congeners in the spirit. They form a chemical equilibrium with their component carboxylic acids and alcohols, but during fermentation they are a by-product of coenzyme A (CoA-SH) recycling when acyl CoA is produced but not required for lipid and protein synthesis (Figure 3.8). Ethanol and acetate are the most abundant acid and alcohol present during fermentation. Consequently, ethyl acetate is the most abundant ester. However, it is not the most important in terms of flavour, as higher esters are more flavour-active and have a greater impact on the flavour attributes of the new make spirit and the final whisky.

The most flavour-active organic compounds contain one or more sulphur atoms. In the context of whisky production, the main contributors to the production of sulphur-derived flavour compounds are the biosynthesis of the sulphur amino acids, methionine and cysteine, and the reduction of sulphate and sulphite in the

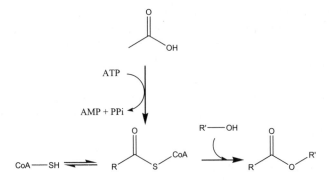

Figure 3.8 Recycling of coenzyme A (CoASH) and its role in the biosynthesis of esters. Acetyl coenzyme A is essential for the biosynthesis of lipids and proteins.

wort. Unlike beer production, the whisky industry has kept faith with copper in their stills, which can partially, but not completely, remove sulphur volatiles in the distilled spirit. (As we will see in Chapter 5, some sulphur compounds can also be lost during the protracted maturation process.)

Finally, it is worth considering the pivotal role of oxygen during fermentation. Yeast cells require oxygen for the synthesis of unsaturated fatty acids and sterols that are essential for the development of cell membranes. This limits growth under anaerobic conditions unless the wort is supplemented with these components directly. However, because of a phenomenon known as the Crabtree effect, when there is appreciable sugar levels present, fermenting yeasts convert sugars anaerobically, whilst the yeast can use dissolved oxygen to grow even when aerobic conditions no longer prevail. Some work has suggested that oxygen helps to restore mitochondrial function and, in principle, such yeasts could perform satisfactorily even without wort oxygenation. Nevertheless, this is likely to influence the flavour profile of the resulting spirit and final whisky.

3.3.2 Distillery Yeast

Distillery yeast is purchased from any of a number of suppliers. It is typically grown up on molasses supplemented with ammonium salts to provide a nitrogen source. It is important to achieve growth aerobically, so the sugar content of the medium should be kept low to avoid the Crabtree effect. Propagation is continued until it is not possible to maintain aerobic conditions. The yeast is recovered by rotary vacuum filtration and supplied in one of three forms: cream (18% dry weight), compressed yeast (24–30% dry weight) and, more recently, dried yeast (92–95% dry weight). The former two need to be kept cool and used within three weeks, whilst dried yeast can be kept at ambient temperatures for up to two years. However, dried yeast requires careful rehydration to ensure that its viability is not unduly lost.

3.3.3 Distilling Fermentations

The fermentation vessels used in the distilling industry tend to be wooden—Oregon pine or larch are common materials of construction—or of stainless steel. The selection of these specific woods is due to the relative lack of branching along the trunk, which in turn means that there are less knots in the fermenter staves. Wooden vessels are used for the fermentation of malt wash, but they are very difficult to keep sterile. They tend to be roughly cylindrical, with volumes of 20–60 000 L and covered by a loose-fitting top. For some distillers, the sterility issue has led to the adoption of stainless steel for the washback construction, as used by grain distillers. Grain distillery washbacks are much larger and can be up to 500 000 L.

3.4 LIQUID–LIQUID SEPARATION: ETHANOL (AND OTHERS) RECOVERY

The process of distillation to separate liquids, or liquids from solids, has been known since antiquity. (Indeed the word 'alcohol' is widely considered to be derived from the Arabic al-kuhl[†].) Relying on the different volatilities of liquids at given pressures, the composition of a vapour above a liquid mixture will have a disproportionately higher concentration of the more volatile component. For the purposes of whisky distillation, several points are worth noting:

1. A substantial proportion of the water introduced into the process to allow saccharification and fermentation must be removed

---

[†]This literally means "*finely powdered antimony*", but is used to indicate the essence of things, including alcohol from fermentations.

2. Ethanol should be recovered, essentially quantitatively
3. The profile of volatiles and semi-volatiles should be 'shaped' according to the requirements for the specific spirit

Clearly, a critical stage in the manufacture of whisky is the increase in the concentration of ethanol. Historically, this has been achieved by the distillation of an alcohol-rich stream from the fermented wash, and is reliant on variations in volatility of ethanol, water and the various congeners present. As mentioned previously, ethanol–water mixtures are non-ideal, which has significant implications for their separation by distillation.

Today, whisky distillation is considered to be either a continuous or a batch process. In Scotland, continuous processes are synonymous with the production of grain whisky, whilst batch processes are generally used for the production of all-malt whisky. The typical configuration of a continuous still to separate binary mixtures relies on multiple points (plates) of liquid–vapour separation (Figure 3.9). As a general principle for ideal, homogeneous binary systems, the closer the components of the binary mixture are in terms of boiling point, the more plates are required for effective separation. At each plate, vapour travelling upwards bubbles into the liquid held up on the plate above and condenses. This results in the release of energy due to the heat evolved in vapour condensation and the evolved heat causing the evaporation of vapour enriched in the more volatile component. When the conditions of a dynamic equilibrium apply, the effect is that the stream of vapour rising up through the plates becomes increasingly enriched in the more volatile component, and the less volatile component effectively moves down the column (Figure 3.10). This stripped liquid leaves the column and is re-boiled, driving off much of the residual higher volatility component and maintaining the heat energy in the column. The less volatile component is then discharged.

However, this general model is not appropriate for the preparation of grain whisky. The feedstock, (fermented wash) is not a binary mixture, but rather contains hundreds of volatile components and a substantial solid loading. One consequence of this is that an ethanol-rich stream is not recovered from the top of the column, but close to the top, allowing compounds of higher volatility (*e.g.*, methanol and acetaldehyde) to be drawn off the top of the column. A distillate withdrawn from the top of the column

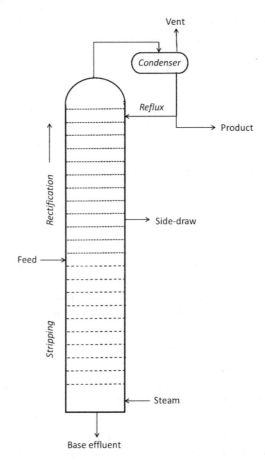

Figure 3.9   General features of a continuous still, highlighting stripping (recovery of volatile components), rectification (concentration of volatile components) and reflux (the partial return of distillate to the column for increased rectification.

may also exceed 94.8% ABV, which is the legal maximum ethanol concentration for Scotch whisky production. In practice, whilst the number of theoretical plates can be calculated, more are needed to compensate for deviations from what is a theoretical minimum. Indeed, as wash typically contains less than 3 mol% of ethanol, the number of plates required for effective separation precludes the practical fabrication of such a still in a single column. Consequently, in practice, a two-column arrangement is preferred. Finally, the use of a reboiler for grain whisky production is

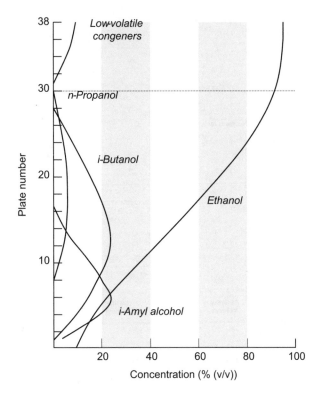

Figure 3.10 Typical distribution of congeners over a rectifying column still. Reproduced from Ref. 11.

impractical as the solid loading in the spent wash is such that unacceptable levels of flavours—predominantly derived from the Maillard reaction—would be generated. In practice, steam is injected to maintain a thermal equilibrium in the still.

A common continuous still configuration is that based on the Patent or Coffey still (Figure 3.11), which is based on an original design by Aeneas Coffey in around 1830. The still is constructed around two columns: an analyser and a rectifier. The analyser effectively strips an alcohol-rich vapour (hot spirit vapour, HSV) from the wash and recovers the spent wash for processing, for instance, for cattle feed, whilst the rectifier concentrates the HSV up to collection strength (Figure 3.12). Cold wash, typically around 30 °C is fed from a washback or, commonly, a wash charger, into a copper coil and flows top-to-bottom through the rectifier. This elevates the wash temperature to 90 °C or more, and is then fed on

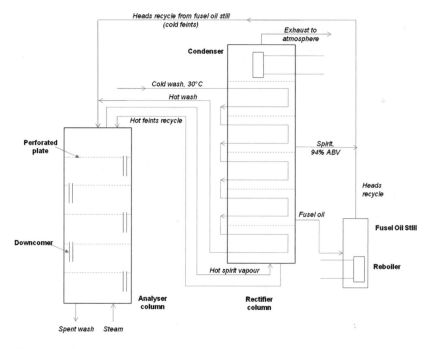

Figure 3.11   Typical liquid flows in a Coffey still.

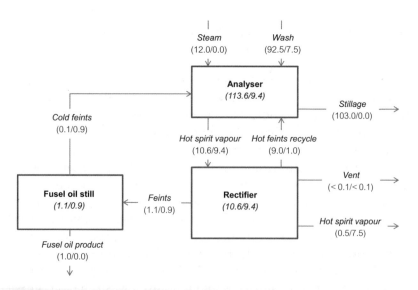

Figure 3.12   Volume fluxes per unit time in a typical grain distillery. Numbers in parentheses represent relative volumes per unit time of water and ethanol, respectively.
Based on data published by Ref. 8.

to the top plate of the analyser column. The hot wash percolates through the plates and downcomers in the analyser, whilst steam passed into the bottom of the analyser distills HSV from the hot wash. The HSV is then fed into the base of the rectifier. Alcohol is consistently concentrated as it passes through the plates in the rectifier and, typically, there are four fractions collected:

1. Volatiles from the top of the rectifier
2. Spirits, typically at 94% ABV, from a plate near to, but not at the top of, the column
3. Fusel oils, which are re-boiled and the vapour fed back into the top of the analyser
4. Hot feints from the base of the column, which again are fed back into the top of the analyser

The recycled flows allow for minimum losses of alcohol, whilst creating a stream of alcohol of sufficient purity for grain whisky production. For the production of grain neutral spirit, such as that used in gin or vodka production, additional, more intricate separation of residual congeners is required.

To ensure that the still runs continuously and predictably, it is essential that the feedstock and energy inputs are balanced. Wash production is a batch process; often a wash charger will hold more than one wash, to help even out the ethanol concentration in the feed. However, traditionally, warm water (30 °C) has been used to adjust the wash to the lowest concentration typically reached during fermentation.

Whilst the majority of Scotch whisky spirit is made in the seven continuous distilleries in Scotland (55.4% across seven distilleries), there were, as of 2008, some 98 malt distilleries producing some 44.6% of all Scotch whisky spirit. Thus, batch distilling tends to be of a much smaller scale than continuous distilling. With the exception of the triple distilling practiced at the Auchentoshan and Springbank distilleries, malt whiskies in Scotland are produced using two pot stills: a wash still and a spirit still. A typical layout of a malt whisky distillery (Figure 3.13) shows that there is a backward propagation of feed in the form of foreshots and feints from the spirit still to be recycled through subsequent wash distillations. The wash is often pre-heated in a heat exchanger against pot ale from the wash still, both to recover energy from the pot ale waste stream and to minimize the risk of charring in the still due to the

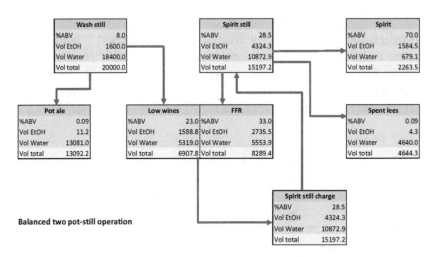

Figure 3.13 Water and ethanol fluxes in a typical malt distillery, based on a wash still charge of 20 000 L. All volumes are given in litres.

application of heat to the cold wash. The wash is often at least 48 hours old, which reduces the risk of overfoaming in the still, which would in turn contaminate the Lyne arm and condenser. Frothing, observed through sight glasses in the wash still, is used as a guide to adjust the amount of heat applied.

The progress of wash distillation is monitored by the use of hydrometry. The initial alcohol concentration coming off the still depends on the alcohol content of the wash, but is often around 45% ABV. Distillation is allowed to continue until the concentration has dropped to around 1% ABV to give a low wines fraction that is around 23% ABV. Further efforts to distill residual alcohol are not considered to be economic in terms of energy required or the time it would take. The residual liquid in the still, pot ale, is typically around two-thirds of the total still charge and generally contains less than 0.1% ABV. This pot ale is often concentrated to a syrup and mixed with draff for disposal, for instance as cattle feed.

The low wines are collected in a buffering tank, termed either the low wines receiver or low wines/feints receiver (LWFR), from which the feed for the spirit still is withdrawn. The feed is generally not preheated, owing to the risk of alcohol loss, which is financially punitive for the distiller. The non-distilled residue from the spirit still, the spent lees, can also be used for the purpose of heat recovery. The spirit still yields three fractions: foreshots, spirits and feints (Figure 3.14). The foreshots are the initial fraction off the still, which serves a dual purpose of cleansing the still of residual high boiling compounds from the internal surfaces of the still left from the previous run and removal of unwanted volatile

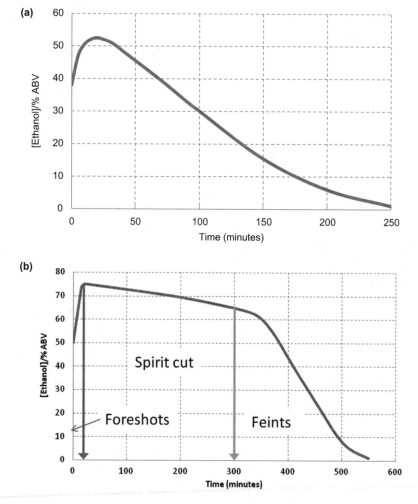

Figure 3.14 Representative development of alcohol strengths during operation of (a) a wash still and (b) a spirit still. Time axes and spirit/feints cuts are highly dependent on the distillery.

components. To make the judgment as to whether sufficient fore-shots have been collected, a demisting test is traditionally performed, although some distillers now base the first cut on the basis of time. For the demisting test, samples of the foreshots are diluted to 47.5% ABV with water. If a cloudy suspension results—somewhat similar to pastis with water—collection of foreshots is continued until such dilution yields a clear liquid (see Figure 4.2 in Chapter 4). The flow from the spirit still is then diverted to collect the spirit fraction at what is known as the first cut, and typically starts at around 73% ABV. Spirit collection continues up to a point, which is defined by the alcohol concentration of the distillate. At some point, often around 61% ABV, the flow is diverted into feints until the alcohol content decreases to about 1% ABV, when distillation is stopped. The exact positioning of the cut-points is often brand-specific and, for peaty malts, spirit cuts may be less than 60% ABV to enhance the concentration of the relatively non-volatile phenols and guaiacols in the final spirit. The residual spent lees will be <0.1% ABV.

Intuitively, it may perhaps be considered that an undue amount of alcohol is distilled as feints (typically 20–25%). However, distillers take great care in the preparation of the foreshots and spirit, as too-rapid collection of spirit gives rise to a fiery, unbalanced spirit and excessive inclusion of higher boiling compounds in the spirit will have an adverse effect on spirit quality. Thus, the feints contain a substantial proportion of the alcohol recovered and must, therefore, be recycled to prevent losses of alcohol. Foreshots and feints are therefore collected either separately or in combination with low wines in the LWFR. In the former situation, the various streams are blended prior to charging of the spirit still; whilst, in the case of the latter, feed is taken from the LWFR as is. In either case, it is essential that the overall alcohol concentration going into the spirit still does not exceed 30% ABV. In the case of the LWFR, it acts as a separating vessel with a separation of heavy oils and higher esters on top of the aqueous alcohol layer. If alcohol contents exceed 30% ABV, this layer starts to solubilise and can result in a so-called blank run, where the distillate from the spirit still never passes the demisting test. However, even if the feed is less than 30% ABV, blank runs can also occur if the surface layer is inadvertently drawn off as feed for the spirit still, for instance by complete emptying of the tank. If the alcohol concentration in the

LWFR approaches 30% ABV, it is often prudent to dilute with water to minimise the risk of blank runs and contamination that can take several subsequent runs to purge the still clean.

For consistent spirit quality, it is important that the spirit cut is consistent so that, in turn, the feed needs to be consistent. It is increasingly common today for malt distilleries to operate in a balanced mode. This means that the feeds into, and the withdrawals out of the receivers follow a regular pattern, both in terms of timing and alcohol strength (Figure 3.15). Thus, the phasing of inputs and outputs should be reproducible, establishing a periodic cycle. Furthermore, phasing with the time cycles of milling, mashing and fermentation will also help to achieve balance. Many distilleries have more than one wash and spirit still, so an additional layer of complexity is the relative phasing of the operation of multiple stills.

3.5   FILLING THE CASK

Before interring new make spirit in the cask, it is essential that both the spirit and the cask are within specification. For the spirit, the major quality criterion is the concentration of ethanol. (We will see in Chapter 4 how the ethanol concentration can affect the progress of whisky maturation in the cask.) The final strength of spirit from the distilling process varies from around 68–71% ABV (for malt distilleries) to over 94% ABV (for grain distilleries). Traditionally, spirit was diluted to a maturation strength of 63.5% ABV prior to cask filling, although some distillers are now maturing Scotch whisky at collection strength. This has the advantage of saving the costs of cask inventory, so maturation at 71% ABV compared to 63.5% ABV reduces cask volume requirements by 12%. Generally, no Scotch whiskies are matured at >80% ABV, with the higher strengths generally reserved for grain whiskies. Higher strengths should be used judiciously, as they will increase the levels of wood lipids and ethanol-soluble lignins, both of which will adversely affect the final processing of the matured whisky, especially in terms of filtration. With the exception of maturing new make spirit at still strength, new make spirit must be accurately mixed with diluting water and, given that such mixing is monitored by density measurements, it is essential that dilution occurs at known

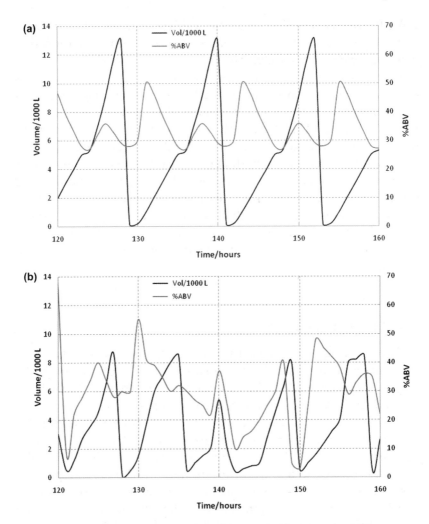

Figure 3.15   Graphical representation of (a) a balanced malt distillery operation and (b) an unbalanced malt distillery operation. The balanced operation is characterised by the repeated pattern of both
volumes and alcohol by volume, whilst the unbalanced operation
demonstrates more chaotic behaviour.

temperatures to ensure that the necessary temperature corrections
are applied.

It is also essential to ensure the quality of casks prior to filling,
particularly to confirm the absence of taints and off-flavours.
These are most conveniently detected by sniffing the cask. Their

presence may indicate a fungal or bacterial infection. Although they will not survive the high alcohol strength of the spirit, the tainting species will be effectively extracted and confer adverse flavour qualities on the final product. Once the cask is charged, it is closed by insertion of a tightly-fitting bung and is ready for transportation to bond to begin maturation.

3.6 CO-PRODUCTS

The production of distilled spirits results in the simultaneous production of various waste streams. Thus, production of spirit for Scotch whisky production yields predominantly liquid and also wet solid streams. As the industry has developed and grown, the effective use, rather than disposal, of these streams has become more important from sustainability and value-added perspectives. This is reflected in the widespread adoption of the term 'co-product' rather than waste stream. A major focus is on the conversion of these co-products into energy sources to reduce reliance on non-renewable energy sources.

Overall, the challenges for managing co-products include:

- An absolute necessity to deal with them in some manner. Whether the co-product is water- or solid-rich, it cannot be stored for a protracted period because of the sheer volumes involved and their propensity to decay
- The often stringent limits concerning the quality of water returned to the external environment, both in terms of composition and temperature
- Ensuring that co-products for animal feed are regularly checked for potentially damaging contaminants, to ensure that they are not passed into the food chain
- Firmly coupling the co-product processing operations to the main business of spirit production in terms of processing rates and coping with variations in production

The common by-products from whisky production (Table 3.10) are quite distinct in terms of their solids and compositional content. The pot ale from malt spirit production is relatively acidic and therefore dissolves some copper from the still. The levels of copper are significant, with a solid content of around 4% (*w/w*) and

Table 3.10 Major co-products originating from malt and grain spirit
production.

	Source	Applications
Malt spirit production		
Draff	Ex-mash tun	Animal feed, dark grains, energy production
Pot ale	Ex-wash still	Pot ale syrup (cattle, pig feed), dark grains
Spent lees	Ex-spirit still	Limited; treated by conventional effluent treatment
Grain spirit production		
Spent wash centrate	Ex-centrifuge or filter	Disposal to sea or concentrated as syrup
Spent wash solids	Ex-centrifuge or filter	Blend with centrate syrup; animal feed

typically 100 mg kg^{-1} dry matter copper. In 2011, total malt spirit
production was 239.3 million litres of pure alcohol so, assuming an
average wash alcohol content of 8% ABV, around 2800 million
litres of wash are produced and around two thirds of this, 1800
million litres, remains as pot ale. This is in excess of the fertiliser or
pig feed needs, so whilst some is still discharged into the sea *via*
deep sea outfalls, most is concentrated to give pot ale syrup (PAS).
This has a solid content of around 50% and is superficially similar
in viscosity and appearance to molasses. Whilst this material can be
used for cattle feed, the market is still rather limited and is preferentially blended with draff to give dark grains. The draff used for
dark grains production is dewatered in simple rotary screw presses
before mixing with PAS, drying and pelleting. Typically, the resulting pellets are modestly friable and have the colour of ground
coffee.

In contrast to the rather similar processes used in malt spirit
production for handling co-products, the various grain distilleries
each have their own bespoke processes and, as a consequence, the
approach to managing co-products varies accordingly. Some grain
distillers remove the mash solids prior to fermentation, whilst in
other situations the mash solids are retained. In either case the
contents of the washback are passed to the continuous still for
alcohol recovery. The spent wash solids are recovered either by
centrifugation or filtration. The filtrate or centrate are ideally

discharged into the sea or evaporated to produce a spent wash syrup. The moist solids, with the addition of a preservative, can be sold directly, or mixed with spent wash syrup to produce dark grains. In 2011 the total production of grain spirit was 316.4 million litres of pure alcohol, so again assuming 8% ABV wash, the centrate/filtrate volumes will be around 3700 million litres. This emphasises the value of having a deep sea discharge option. An alternative way of looking at such data is to assume the situation of a large—100 million lpa —grain distillery. Such a distillery will produce around 1200 million litres of filtrate/centrate, or 24 million litres (*ca* 24 000 tonnes) per week.

For malt distilleries, the production of PAS and dark grains is challenging, not least on the basis of meeting ever more stringent environmental standards to the extent that it has become uneconomic for the smaller distilleries to process their own co-products, resulting in more centralisation of co-product production at larger plants. The situation for grain distilleries is rather more straightforward, with scales from 10–100 million lpa justifying the co-product treatment capacity.

Whilst it is feasible to recover value from draff, spent wash and pot ale, there are other waste streams that have no value, which nonetheless need to be dealt with to ensure compliance with current legislation. Such streams include spent lees, washing waters and condensates from PAS and dark grains production. The routes for the treatment and disposal of these streams depend on the location of the distillery and, in Scotland, are defined by various pieces of UK and EU legislation. The Scottish Environmental Protection Agency (SEPA) have set a range of environmental standards, with temperature, dissolved copper and dissolved oxygen being particularly relevant to the operation of Scottish distilleries.

The levels of copper in pot ale is lower than in spent lees, but in the former the copper is mainly bound into the solids, whilst in the latter it is essentially solvated copper(II). The copper levels in sludge from spent lees and wastewater can elevate the copper levels in soils on to which it is sprayed. This is beneficial in Scotland, given much of the land is deficient in copper, but of course requires monitoring to ensure legislative compliance. Other options for copper removal from spent lees have also been actively explored. As much copper is present as soluble copper, it can be precipitated, deposited electrolytically or ultrafiltered to reduce the copper effluent burden. The electrolytic approach has the advantage that copper can be recovered as copper metal.

The markets for co-products have historically been for animal feed, fertiliser and fuel, with small potential outlets as human food and biomass production. Today, much effort is focused on the development of energy streams. The feed industry is dominated by ruminants (cattle and sheep) as they are able to cope better with the cellulosic content of the co-products. Other animals, particularly pigs and poultry, can be fed co-products to a certain extent, but they lack some essential amino acids, with low lysine levels being the most problematic.

Feeds could, in principle, be used as direct fuels, but their value as feed would have to be very low to make this a viable option. Some efforts have been made to mix moist co-products with wood chips as an energy source, but again these approaches tend to be short-lived. Much more promising are the latest generation of bioenergy plants. For instance, Diageo have invested some £65 million in their bioenergy plant at its grain distillery at Cameronbridge. The plant includes anaerobic digestion, biomass conversion and the application of ultrafiltration and reverse osmosis to concentrate dilute aqueous feeds. The principle is to remove solid waste for burning to generate heat and energy. The resulting liquid stream is treated in an anaerobic digester to generate biogas as fuel for boilers. It is expected that the combined energy production will satisfy some 80% of electrical and 98% of steam needed in the distillery, as well as addressing the requirements for effluent treatment. Similarly, a consortium of Rothes distillers in Speyside, together with an energy company, have teamed up to construct a combined heat and power plant based on distillery biomass, with energy outputs that are expected to be sufficient to power 9000 homes.

Whilst it is still too early to say whether such plants will realise their full potential, their development suggests innovations that can be achieved as technology catches up with the ambition of the industry. This is exemplified by the highly-publicised environmental strategy of the Scotch Whisky Association, which includes a commitment to reduce the reliance of the Scotch whisky industry on fossil fuels in the future.

3.7 IN CONCLUSION...

The initial stages of whisky production, converting cereals and extracts into fermentable substrates, and their subsequent fermentation and distillation, takes relatively little time to complete, with crop-to-cask requiring 2–3 weeks of processing to complete. Each operation of malting, mashing, fermentation and distillation have been refined over years of production. These processes are of critical importance both for setting the scene for final product quality and the incurred costs that, with the exception of income derived from the sale of new make spirit or from co-products, are not recovered until the final product sale years later.

REFERENCES

1. M. D'Angelo, G. Onori and A. Santucci, *J. Chem. Phys.*, 1994, **100**, 3107–3112.
2. M. Macatelli, A. Paterson and J. R. Piggott, Spirit flavour release under mouth conditions, In: *Proceedings of the Second Worldwide Distilled Spirits Conference*, Nottingham University Press, Nottingham, pp. 151–158.
3. D. A. Nicol, *Batch distillation, In: Whisky - Production, Technology and Marketing*, ed. I. Russell, Elsevier, Cambridge, UK, 2003, pp. 153–176.
4. M. C. Meilgaard, *Tech. Q. Master Brew. Assoc. A.*, 1975a, **12**, 107–112.
5. M. C. Meilgaard, *Tech. Q. Master Brew. Assoc. A.*, 1975b, **12**, 151–158.
6. D. E. Briggs, *Malts and Malting*, Blackie Academic & Professional, London, 1998.
7. L. Nykanen and H. Suomalainen, *Aroma of Beer*, Wine and Distilled Beverages, Springer, London, UK, 1983.
8. I. Campbell, Grain whisky distillation, In: *Whisky - Production, Technology and Marketing*, ed. I. Russell, Elsevier, Cambridge, UK, 2003, pp. 179–206.
9. C. Boulton and D. Quain, *Brewing Yeast and Fermentation*, Wiley-Blackwell, Oxford, UK, 2006.
10. M. Jones and J. M. Pierce, *J. Inst. Brewing*, 1964, **70**, 307–315.
11. B. R. Whitby, Fermented Beverage Production, 1992, p. 261.

CHAPTER 4

Wood Chemistry and the Maturation of Whisky

4.1 INTRODUCTION – WHY MATURATION?

It is commonly accepted that the process of whisky maturation was serendipitous, which is not unreasonable given that the oak cask has been the most common storage vessel for liquids probably since the Roman era until modern times. (The idea of forming a barrel shape from staves seems to have been known from between the XVIII and XXVI Egyptian dynasties.)[1] The ubiquity of the cask for storage is evident from historical records. For instance, Henry VIII set several legal capacities of cask by legal statute, the size of which was dependent on the product to be stored (*e.g.*, oil, tar, pork, wine, vinegar) and geographical location. Thus, the London ale barrel was 32 gallons and the country ale and beer barrel was 34 gallons. George III made 38 gallon casks the legal measure for fish and 50 gallons for salt fish. Today, this plethora of barrel 'standards' still exists to some extent (Table 4.1) but now, as then, their one commonality was that they were coopered containers.

The design of the oak cask (Figure 4.1), a double arch structure, is astoundingly strong and also enables even full casks to be moved relatively easily. Indeed, the success of this design, developed more

The Science and Commerce of Whisky
By Ian Buxton and Paul S. Hughes
© Buxton & Hughes, 2014
Published by the Royal Society of Chemistry, www.rsc.org

Table 4.1 Common cask sizes used in the Scotch whisky industry.

Name	Volume/L	Sources
ASB[a]	191	Bourbon industry
Dumphogshead	254	Reworked ASBs
Butt	500	Ex-Spain
Puncheon[b]	558	Ex-UK

[a]American standard barrel.
[b]Uncommon today. Volumes can vary.

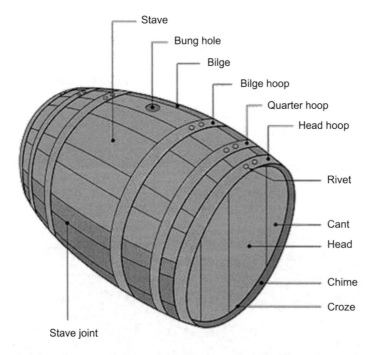

Figure 4.1 Structural elements and terminology of the oak cask.

than 2000 years ago is reflected by the fact that it has not been significantly enhanced since. It is likely that most whisky was historically consumed unmatured, a practice that prevailed until recently in some distilleries with the practice of 'dramming', giving new spirit, or clearic, to distillery staff. Nevertheless, it was recognized that storage of spirit in oak casks fortuitously improved the sensory (*i.e.*, visual and taste) qualities of whisky. Although there has been a recent trend from some distillers to market new make (*i.e.*, unmatured) spirit, the additional costs and time taken to produce matured spirit are clearly valued from a consumer perspective, as shown by the year-on-year global sale volumes and the efforts made to clearly define whisky, particularly in Scotland and, as a corollary, across the EU.

The chemistry of whisky maturation is complex. It is dependent on the history of the cask, the composition of new make spirits and the ambient conditions that prevail during the time course of maturation. Also, there are constraints on maturation that are imposed by the definitions and provenance of specific products. Thus products, such as American bourbons or rye whiskies, are matured in new, charred casks (Figure 4.2), while Scottish and Irish whiskies are matured in previously used bourbon or sherry casks. Canadian whiskies are matured in a wide range of previously-used casks.

An analysis of new make spirit and final matured spirit does not show any gross differences in composition (Table 4.2). Essentially there are no non-volatile components in new make spirit, with the exception of any copper picked up during the condensation of the spirit and ions present in water used to adjust the ethanol content

Figure 4.2 Cross-section through a typical bourbon cask.
(Reproduced with permission from Brown–Forman Cooperage, Louisville, KY.)

Table 4.2 Typical composition matured grain and (unpeated) malt spirits.

Analyte[a]	Sources	Matured grain	Matured malt
Major analytes (g/100 L pure alcohol)			
Acetaldehyde	Fermentation, ethanol oxidation	6	12
Ethyl acetate	Fermentation, distillation, maturation	20	25
Acetal	Ethanol/acetaldehyde reaction	1	4
Methanol	Fermentation, transesterification	10	4
n-Propanol	Fermentation	75	50
Isobutanol	Fermentation	40	70
Isoamyl acetate	Fermentation	1	2
n-Butanol	Fermentation	0.0	1.5
2-Methylbutan-1-ol	Fermentation	8	40
3-Methylbutan-1-ol	Fermentation	20	120
Diacetyl	Fermentation, lactic acid bacteria	2.5	3.0
Minor analytes (mg/L pure alcohol)			
Ethyl hexanoate	Fermentation	0	2
Ethyl octanoate	Fermentation	1	10
Ethyl decanoate	Fermentation	2	9
Ethyl dodecanoate	Fermentation	2	10
2-Phenylethyl acetate	Fermentation	0.25	2.5
2-Phenylethyl alcohol	Fermentation	2	7
Furfural	Heating of sugars	1	4
Gallic acid	Wood-derived	2.5	1
Ellagic acid	Wood-derived	5	3
Coniferaldehyde	Wood-derived	0.5	0.5
Vanillin	Wood-derived	1	1
Vanillic acid	Wood-derived	0.5	0.5
Sinapaldehyde	Wood-derived	0.5	0.7
Syringaldehyde	Wood-derived	2.5	2.5
Scopoletin	Wood-derived	0.3	0.3
5-Hydroxymethylfurfural	Spirit caramel	0.5	0.2

[a]Concentration units reflect common industry practice.

of the spirit prior to maturation. Matured spirit will, however, contain non-volatile substances from the oak wood itself, which will be discussed further below. However, whilst in purely gravimetric terms the differences in the composition of new make and matured spirit are seemingly minor, they belie the flavour activity of compounds that are present at low concentrations, such as dimethyl sulfide (DMS), dimethyl trisulfide (DMTS) and acrolein, as discussed in chapter 3. One of the most striking differences though is in the visual presentations, with new make spirit being colourless and matured spirit being decidedly amber or brown. As we shall see below, this is due to the extraction of colour from the walls of the cask, which are derived from the structure of the wood itself and from the spirit previously held in the cask. In this chapter, we explore these changes that occur over a period of months and years, which are primarily chemical, the high ethanol content essentially precluding any microbial action within the cask throughout the maturation period.

The requirement for maturation, then, is primarily to satisfy the requirements of various pieces of legislation, which in themselves are made worthwhile by the clear enhancements to the spirit that maturation brings about. This is in contrast to other spirit categories, such as tequila and rum, where maturation is an option. Today, there is an increasing trend to focus on maturation of other spirit categories, with the one recent extension being the development of extra añejo (ultra-aged) tequila.

4.2 OAK AND CASKS

4.2.1 Cask Requirements and Supply

For many of the world's whiskies, there is no option other than to expose spirit to oak for a period of months or years. To satisfy legal requirements, this means placing the spirit into oak casks (putting spirit in the wood) rather than putting wood into the spirit. This latter option may well be tenable, but generally falls outside most of the regulations dictating the production of whisky. This reliance on casks adds substantially to the cost of whisky production, not only in terms of purchasing casks, but also because there is a need to manage and maintain cask inventories over a period of years. Most Scotch whisky is matured in American oak casks that have been used for bourbon production. Some inventory is also made up of sherry butts from Spain. The ex-bourbon barrels are typically 190 L (*ca* 50 US gallons) and traditionally these were broken down into their component staves or 'shooks' prior to transportation, then reassembled in Scotland with extra staves and new cask ends to create 250 L hogsheads. Sherry butts (500 L) are highly prized in the Scotch whisky industry. They tend to be longer lasting and impart desirable flavour and colour characteristics to the spirit. Nevertheless, there is a mismatch between the volume of sherry shipped in casks *vs* the volume of whisky produced, such that, despite their undoubted qualities, the US will be the primary source of casks for the foreseeable future.

Because of the relative scarcity of authentic sherry butts, several methods have been developed in an attempt to mimic their performance. Essentially, there are three options available: sherry treatment itself (*i.e.*, production of authentic sherry casks), wine treatment or steam/ammonia (Gomez) treatment. The traditional sherry cask was, strictly speaking, a 500 litre cask used for shipping sherry, an application that followed on from the cask's original use for wine seasoning. Thus, the contact time of the sherry with the cask was typically of the order of 6–9 months and indeed there were casks available until around the mid-1960s. Subsequently, supply diminished and to make up for the shortfall in sherry cask supply alternatives were sought. The so-called Gomez treatment was the application of ammonia and steam under pressure to an American oak cask, with around a litre of 880 ammonia[2] applied at 8 psig for an hour. The main action was to strip tannins from the interior of the cask and the resulting spirit tended to be light in colour, low in terms of extractives from the wood and the maturation of the contained spirit was rather slow.

Wine seasoning was developed to try to mimic both cask seasoning and the hydrolytic effect of sherry contact with the wood. Here, casks are manufactured in Spain, either from Spanish or American oak. Whilst they are made specifically for Scotch whisky, they are used initially for grape fermentations and then exposed to sherry for varying periods before being shipped empty to the UK. An interesting modification of the wine seasoning approach was the development of pressure treatment with a blended sweet, dark sherry known as Paxarette. This is generally very highly coloured and very sweet. Typically, Paxarette is added at a rate of 2 mL L^{-1} cask volume and forced into the wood under pressure (around 7 psig for 10 minutes). The residual Paxarette is disgorged and re-used, and the cask checked again before filling to ensure that there is no excess Paxarette present. This is important as the sheer colour intensity of the Paxarette could easily result in a matured spirit above its colour specification.

4.2.2 Why Oak?

The choice of oak for cask production is due to the fact that it is possible to fabricate casks that are sufficiently impervious to liquid for the duration of maturation. Oak also tends not to contribute undesirable flavour components to maturing spirit. Nevertheless, there is variability between oak varieties and in Europe two varieties – *Quercus sessilis* and *Quercus robur* – are the most commonly used. Sherry casks are made from these species but they are also constructed from imported white oak from the USA. In contrast to Europe, there are a plethora of oak species used for cask production in the USA, with *Quercus alba, Quercus bicolor* and *Quercus microcarpa* being commonly used. In recognition that an oak tree can take 80 years or more to mature before it is harvested, the Scotch Whisky Association has committed itself to make efforts to ensure that oak sourced for Scotch whisky casks comes only from sustainable sources by 2050.[3]

The suitability of oak for cask fabrication is due to two distinct structural features: medullary rays and tylose disruption of xylem vessels. Medullary rays extend from the centre of the trunk out towards the bark and are tough, impervious structures that can contribute more than 25% of oak by volume. It is thought that medullary rays enhance the strength and flexibility of oak, an essential property when it is being used for producing casks. The presence of xylem vessels in sapwood is essential for the transportation of water from the root to the leaf of the tree. As the sapwood turns to heartwood, the xylem vessels, which can be up to 0.3 mm in diameter, are, in certain species of oak, occluded with tyloses, which are thought to develop as a protective mechanism for the maturing tree (Figure 4.3). Certainly tyloses seem to act as a barrier to the translocation of liquids. The composition of tyloses appears to be based on cellulose and is presumably an aspect of wood aging to ensure the robustness of the ever-thickening trunk of the oak.

The structural elements of wood are mainly polymeric, being made up of cell walls (cellulose, hemicelluloses and lignin) with intercellular space consisting predominantly of lignin. Low molecular weight compounds – extractives – make up a relatively small proportion of freshly harvested wood. Cellulose is the main component of wood, accounting for typically 50% by mass.

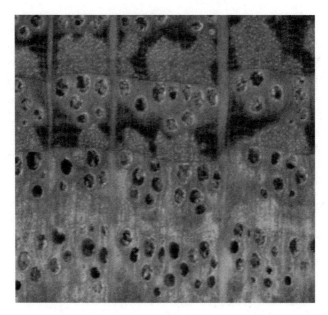

Figure 4.3   Tyloses in the heartwood of white oak indicated by the blocked appearance of the pores.

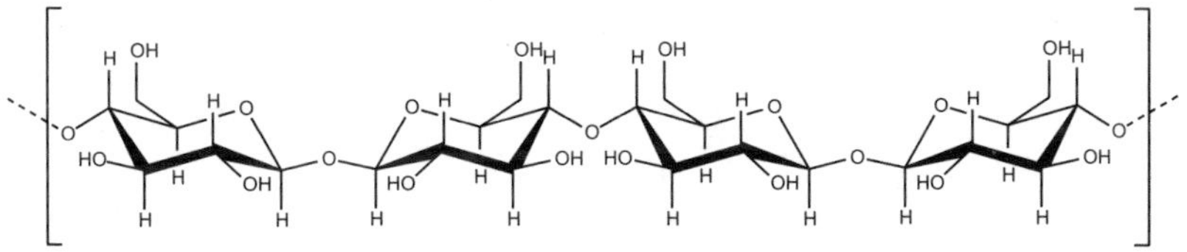

Figure 4.4   A short segment of the cellulose polymer.

Cellulose is a linear polymer made up of β(1–4)-linked glucose molecules (Figure 4.4). The regularity of the cellulose chains allows them to effectively hydrogen bond across chains through exposed hydroxyl groups, giving rise to networks of fibrils that form a framework on which the wood superstructure can be laid down. Hemicelluloses are a diverse group of heteropolymers made up of pentoses, hexoses, hexuronic acids and deoxyhexoses (Figure 4.5). Oak wood hemicelluloses are mainly based on xylose and are typically 15–30% of oak wood by mass. Lignin occurs both in cell walls and in the intercellular space and can be considered to bind the cellular structure together. It is a high molecular weight polymer, made up of monomers based on phenylpropane units.

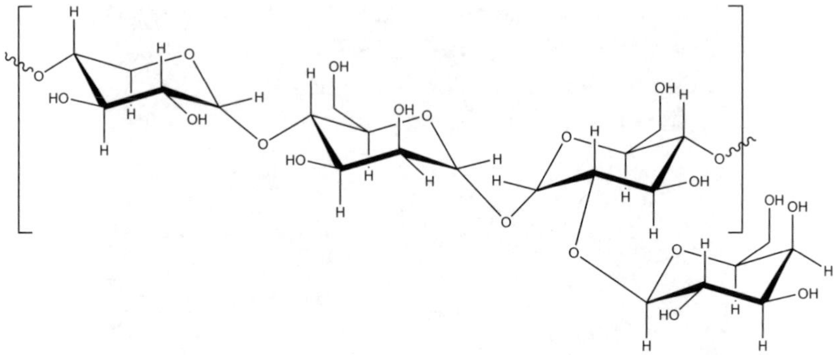

Figure 4.5   A typical fragment of hemicellulose, featuring xylose–mannose–glucose–galactose.

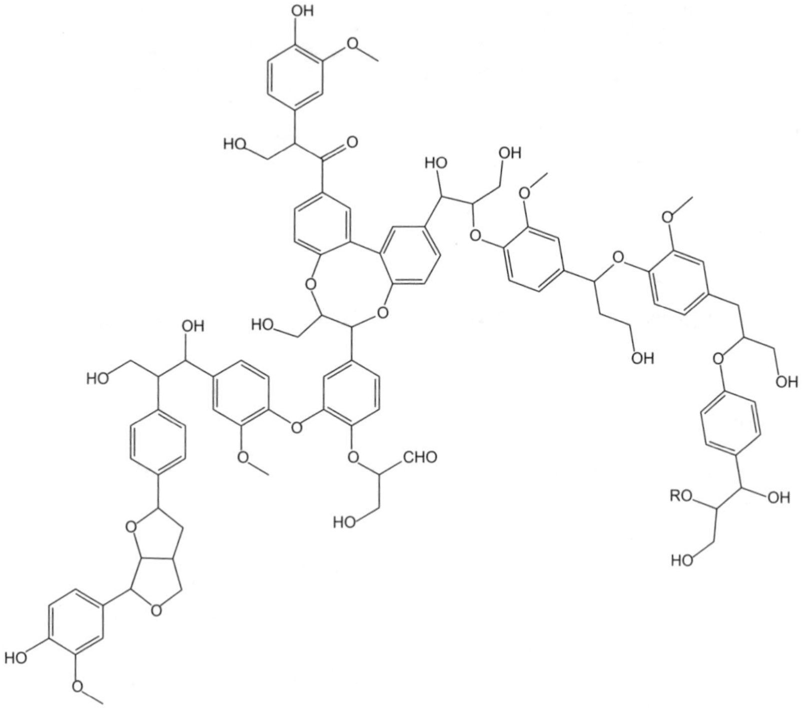

Figure 4.6   A typical fragment of the lignin polymer. R groups indicate a continuation of the lignin polymer.

The aromatic rings are substituted with a range of hydroxyl and methoxyl moieties (Figure 4.6). Of the extractives derived from oak wood, the most relevant to the distiller include the hydrolysable tannins and volatile components. Tannins are well-known for eliciting astringency and bitterness in food systems, and can account for the astringency of oak wood extracts. Various structures based on gallic and ellagic acid have been identified, although heat treatment of oak in the cooperage releases these as free acids, which are prevalent extractives in maturing spirits. The tannins, especially the hydrolysable ellagitannins, are deposited in the lumen of dead wood cells as the sapwood matures into heartwood. This results in a substantial increase in tannin content, and there is typically an order of magnitude difference in the concentration of tannins in sapwood *vs.* heartwood. Of the volatile compounds identified in oak wood extractives, the organic acids, phenols, lactones and norisoprenoids are of most interest. Acetic and linolenic acid (18 : 3) are the major organic acids from oak, the latter being sensitive to oxidation due to the presence of two labile stepped diene functionalities, which in turn can give rise to highly flavour-active oxidation products, such as low molecular weight aldehydes. Of the lactones, *cis-* and *trans*-oak lactones (Figure 4.7) are particularly relevant to the flavour of matured spirit.

4.2.3 Cask Fabrication

Mature oak trees are felled and those that are used for cask production are cut into lengths, either for stave or heading lengths, the latter being used for cask ends. The selection is dependent on the features and defects in the logs. In both cases, the oak logs are sawn into 'quarters' and each quarter is sequentially pared for staves until there is insufficient wood for further use (Figure 4.8a). The sapwood, the layer of wood below the bark, is then cut off to provide the finished staves and heading lengths. The sawn wood is unsuitable for cask manufacture as is and needs to be dried

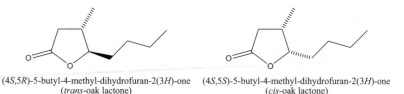

(4*S*,5*R*)-5-butyl-4-methyl-dihydrofuran-2(3*H*)-one (4*S*,5*S*)-5-butyl-4-methyl-dihydrofuran-2(3*H*)-one
        (*trans*-oak lactone)   (*cis*-oak lactone)

Figure 4.7 The *cis-* and *trans*-oak lactones that may be extracted from North American, European and Japanese oaks.

Figure 4.8   a) Wood sawing from bolts; b) a comparison of kiln dried (background) and air-dried staves; c) detail of trimmed staves. Note the
range of widths, reflecting the original bolts from which they were cut.
(Reproduced by permission from Brown–Forman Cooperage,
Louisville, KY.)

(Figure 4.8b). Traditionally, this was achieved by air drying for
several months or even beyond one year. In Spain, timber is usually
dried in the north at the point of harvest for around nine months,
before moving it to the warmer sherry regions in the south for up to
another nine months. Final moisture levels are around 14–16%. In
the USA, oak is either kiln or air dried. Kiln drying takes around
one month, reducing moisture content to around 12%. In any case,
the drying of oak results in wood shrinkage in all dimensions, but
especially along the length of the stave. Undue acceleration of oak
drying risks cracking or splitting of the oak, whilst control of
temperature and humidity minimizes the risks of physical damage.
   Casks for the maturation of distilled spirits are generally known
as tight barrels as they need to be able to retain liquids. This is in
contrast to so-called slack barrels, which were historically produced mainly for the storage of dry goods. The latter require
less stringent coopering than the former. There are a number of
variations around the construction of casks. However, broadly
speaking it proceeds as follows and summarised in Figure 4.9.

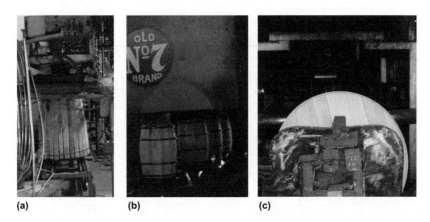

Figure 4.9 The aspects of bourbon cask fabrication: a) raised cask *en route* to steaming prior to being pulled into the cask shape; b) the cask, held in place by iron hoops, in the process of charring; c) a formed, bevelled head about to be charred.
(Reproduced with permission from Brown–Forman Cooperage, Louisville, KY.)

The dried, rough staves are first jointed. Here, the edges of the wood are smoothed, such that the ends of the stave are narrower than the middles and the outside of the stave is somewhat wider than the inside. This sets up the nascent staves to allow the formation of the eventual cask shape (Figure 4.9a). The staves are then set in a form, closed at one end and, with steam conditioning of the staves to make them more pliable, they are drawn together at the open end to form the cask shape. Iron hoops are used to keep the cask shape in place (Figure 4.9b). Most of the stress in the curved staves is at the bilge and, to reduce this stress, the cask form is 'fired'. This is distinct from charring or toasting, and effectively removes the water taken up during stave steaming, with subsequent shrinkage of wood fibres and to set the shapes of the staves so that the retaining hoops are not required for maintaining the cask shape. The body of the cask is then ready for toasting or charring (Figure 4.9c).

The procedures used for producing bourbon and sherry casks are broadly similar, although there are differences (Table 4.3), predominantly in the timing and the degree of heating. Heat treatment of the cask is essential for the maturation of whisky, but the heating applied varies in its severity. Charring is the more

Table 4.3 The typical stages in the production of bourbon and sherry
casks.

Bourbon cask	*Sherry cask*
Shaping, tapering of *ca* 30 staves per cask	Shaping, tapering of *ca* 50 staves per cask
Staves raised into a circular structure	Staves raised into a circular structure
20 minute steaming (95 °C)	Open fire toasting (200 °C), water applied to staves
Staves drawn into barrel shape 1. Heating at 230–260 °C for 15 minutes 2. Charring	Staves drawn into barrel shape
Cask ends fabricated and inserted	Cask ends fabricated and inserted
Hoops driven on for final cask	Hoops driven on for final cask

extreme, with higher temperatures being applied for shorter periods of time compared to toasting, which is a milder but a more
prolonged treatment.

After toasting or charring, the cask needs to be finished off. This
consists of cutting the chime (the bevel at the ends of the staves),
cutting the croze (the grooves into which the heads fit) and cutting
the howel, a slightly rounded cut above and below the croze. The
heads, which may or may not be heat-treated, now need to be
inserted. They are constructed from flat wood, smoothed to the
correct thickness and joined together with dowel pins. After cutting
the heads into circles, their rims are bevelled on both sides and the
heads inserted into the cask body. The form of the casks is kept
tight by driving hoops on to the cask. A bung-hole is then cut,
usually at the bilge and, typically, the cask is pressure tested with
steam *via* the bung-hole to confirm the tightness of the cooperage.
The moisture from this steam keeps the inside of the cask hydrated,
facilitating the maintenance of tight cooperage, at least for a limited period of time.

The adaptation of American standard barrels to hogsheads,
where cask volumes are increased by around 25–30%, brings its
own challenges. One motivation is that the larger casks make better
use of the footprint of a warehouse. Thus, expanding a cask from
190 to 250 L increases cask volume by around 32%, whilst the
footprint of a cask – either palletised on in stow – only increases
by 15%. This represents a significant cost saving for warehousing.

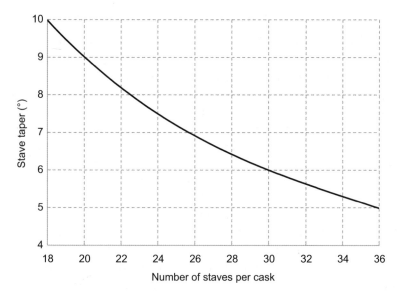

Figure 4.10 The dependence of stave taper on the number of staves employed, the simplest case being when all staves are of equal widths. Increasing a cask from 190 to 250 litres, by adding new staves, is equivalent to adding around one extra stave for every seven present. This is roughly equivalent to a 1° adjustment for each taper.

The angle of the stave taper edges required for optimum bevel-to-bevel contact of cask staves will be dependent on the number of staves employed, the angle being less with the more staves that are inserted (Figure 4.10). Thus, when extra staves are inserted, rather than them butting up against each other by touching surfaces, the joins will tend to be along lines. There are two options available to the cooper, both of which are applied in practice. The staves can be trimmed to account for the slight reduction in the angle of the bevels, so that full bevel-to-bevel contact is restored. Another option is to force the oak into a slightly different geometry. By pulling the staves of the larger, newly-raised casks very tightly together and applying steam to make the wood more pliable, the oak can be deformed to some extent, enhancing the tightness of the cooperage. In either case, the result is a cask that fulfils the requirements of tight cooperage and are both considered to be suitable for whisky maturation. Of course, new heads need to be fabricated for the larger casks.

4.2.4 Heat Treatment of Cask Surfaces

There are three key aims of cask heating: breakdown of wood polymers to yield flavour extractives, thermal degradation of undesired wood-derived flavour components and, for charred casks, to create a surface of active carbon (Figure 4.11). The breakdown of wood polymers generates both colour and flavour compounds. The thermal degradation of sugars and polysaccharides is complex and highly dependent on the local water activity. Hemicellulose, based on five-carbon sugars, is more thermally labile than cellulose, which accounts for the fact that of the furaldehydes formed, furfural predominates. Furaldehydes are not particularly flavour-active in their own right but can be considered to be markers of the thermal degradation process, which also generates other, more flavour-active species. These have caramel, sweet and toasted aromas, which are due to the presence of a whole range of chemical entities, including furaneol and 2,3-dihydromaltol.

Lignin degradation products are potentially highly flavour-active. Vanillin, syringaldehyde, coniferaldehyde and sinapaldehyde can

Figure 4.11   Detail of a freshly charred bourbon cask.
(Reproduced with permission from Brown–Forman Cooperage, Louisville, KY.)

undergo further degradation in both the wood and the maturing spirit to give vanillic and syringic acids. Vanillin is especially important because of a combination of its low threshold and typical levels that arise in the final spirit. Charring of casks increases the levels of lignin breakdown products that can be extracted into spirit, not so much in the char layer itself but in the subsurface layers that are subject to a lower heat load. These compounds are readily extracted as the char layer itself is porous, and splits and cracks when heated. Indeed, a cross-section of a stave, which has been part of a cask used to mature whisky, shows a clear demarcation indicating the extent of liquid penetration into the wood. Of the other components in wood, of particular note are eugenol and the oak lactones. It is not clear whether heat treatment results in a net increase or decrease of these compounds as their volatility is at least partially offset by possible development during cask heating. Aromatic aldehydes are thermally labile and have been shown to occur at lower levels if cask treatment is at higher temperatures or if there is charring, resulting in the formation of volatile phenols, such as guaiacols and syringols (Figure 4.12).

The development of colour is an essential requirement for maturing spirit. Heat treatment results in the increasing development of extractable colour. Colour compounds tend to be of larger molecular weight and their chemical structures have not been fully elucidated, but it is reasonable to assume that there is a significant increase in the levels of highly conjugated species. As molecular weight increases during heating, the resulting species become less soluble, so whilst colour contribution increases with the level of toasting and charring, the char layer itself contains lower concentrations of colour species and, therefore, makes little direct contribution to spirit colour.

The removal of unwanted flavour-active compounds in oak casks should not be underestimated. Unsaturated fatty acids, such as linolenic acid, are prone to oxidative reactions to give highly flavour-active degradation products. Unsaturated aldehydes, such as *trans*-2-octenal, and *trans*-2-nonenal, and the intensely flavour-active 1-octen-3-one, can confer metallic, mushroom, sawdust/woody and cardboard-like flavours. Whilst active carbon makes little direct contribution to the flavour or colour of maturing spirit, it does have a pivotal role to play in the progress of maturation, not least the oxidation of DMS to DMSO (as discussed in the next section).

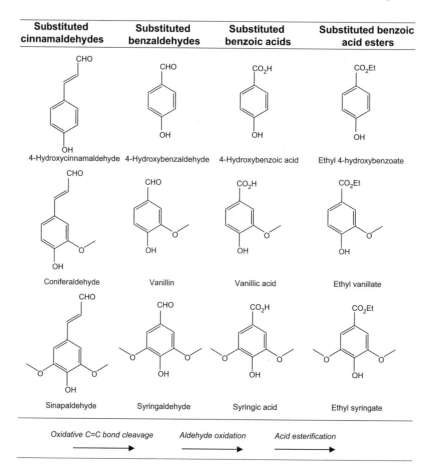

Substituted cinnamaldehydes	Substituted benzaldehydes	Substituted benzoic acids	Substituted benzoic acid esters
4-Hydroxycinnamaldehyde	4-Hydroxybenzaldehyde	4-Hydroxybenzoic acid	Ethyl 4-hydroxybenzoate
Coniferaldehyde	Vanillin	Vanillic acid	Ethyl vanillate
Sinapaldehyde	Syringaldehyde	Syringic acid	Ethyl syringate

Oxidative C=C bond cleavage   Aldehyde oxidation   Acid esterification

Figure 4.12 Typical chemical conversions of lignin-derived aromatic species, the most important of which is the formation of vanillin.

4.2.5 Cask Performance

Once a cask has been used several times for spirit maturation – typically three, four or five times – the cask becomes more and more spent or 'done'. It will typically take a long time before a cask is done, perhaps 40–50 years. The cask can be discarded at this point, although there is increasing interest in the option of re-generating or rejuvenating the cask. There are several ways in which a cask might be at least partially rejuvenated, but they generally follow similar lines, namely wire brushing or routing the inside of the cask to expose fresher wood and then re-charring the surface. Such approaches can replenish several of the cask components, particularly those derived from cellulose, hemicellulose and lignin degradation. However, it is noteworthy that the levels of tannins and the oak lactones are not replenished. Thus, a rejuvenated cask does not behave as a conventional cask should but it does, however, offer options for extending the lifespan of at least part of the cask portfolio, to the point whereby the working life of a cask may exceed the time that it takes to re-grow a matured oak tree. Such casks may be considered to satisfy the SWA requirements for sustainable oak.

Before we conclude the discussion of the cask, it is worth considering two other aspects of the cask that can, in principle, affect the progress of whisky maturation: the impact of cask surface area/volume ratios and the role of stave joins.

Given the importance of the inner surfaces of the casks to the maturation process, it is logical to consider the ratio of the surface area to volume. The surface area of a cask is in units of $(length)^2$, whilst volume is in units of $(length)^3$, so that the ratio always has units of $(length)^{-1}$. Calculations of surface areas, volumes and their ratios for a range of cask sizes (Table 4.4) suggest that maturation can be expected to occur more rapidly in smaller casks, at least in terms of evaporative losses and surface-dependent chemical reactions. Furthermore, the mean diffusion pathway length will be longer for the larger cask. Although this is not straight-forward to calculate for the cask geometry, the equation for Brownian displacement indicates that the distance of diffusion in a given direction is proportional to the square root of time:

$$\bar{x} = \sqrt{2Dt} \qquad (4.1)$$

where $\bar{x}$ is the mean distance of diffusion, D is the diffusion coefficient and t is time.

Table 4.4 Typical surface area/volume ratios of casks, calculated on the basis of Ref. 11.

Name	Volume/L	Surface area/dm²	Surface area : volume/dm⁻¹	Relative surface area : volume
ASB	191	204	1.07	1.37
Dumphogshead	254	261	1.03	1.32
Butt	500	316	0.78	1.00

The combined effects of a smaller surface area to volume ratio and longer mean diffusion paths for larger casks means that maturation will generally take longer to complete in larger casks and, in principle, there may be a shift in the balance of the various mechanisms of maturation, although as yet there have been no reports on this.

The observation that there is a limit to which spirit appears to penetrate into staves begs the question as to whether the migration of spirit through the wood itself can fully explain the evaporative losses from casks. By way of example, consider a cask of 200 L, with a surface area of close to 200 dm^2. It may be expected that, with a 2% volume loss per annum, typical of Scottish maturation conditions, around four L will be lost during the first year of storage, or around 20 mL dm^{-2} of cask surface. This is a significant flux through the cellular structure of oak, the varieties of which have been selected for their multiserate (and, therefore, essentially impervious) medullary rays. Three other routes present themselves:

1. Physical damage to the casks themselves
2. Evaporation through imperfections in the joins between staves
3. Partial diffusion of liquid into the staves and subsequent translocation through non-occluded xylem vessels

Of these, the evaporation of spirit through imperfections in stave joins seems to be the most likely as they will exist, to some extent, in all casks. In the wine industry, it is thought that oxygen ingress increases by a factor of three to four as cask contents evaporate, which implies that losses through the staves in contact with liquid are slower than through the presumably drier uncovered staves. For the creation of hogsheads from American standard barrels, it is interesting to speculate whether the reworking of staves or the forced compression of steam-conditioned staves results in the tighter cooperage and, therefore, reduced evaporative losses.

4.3   WHISKY MATURATION

The strength of new make grain spirit leaving the column still (*ca* 94% ABV) is too high for filling directly into cask. Indeed, the spirit strength from the malt spirit still, at above 70% ABV, is often considered too high, although some distillers are filling casks at higher alcohol concentrations, not least because of the savings in cask inventory. Thus, filling at 71% ABV compared with a more typical 63.5% ABV means that there is a saving of about one cask in every ten. Nevertheless, the solvating properties of ethanol/water mixtures are dependent on the proportions of the two liquids. We might expect that filling at higher fill strengths will favour the extraction of more lipophilic compounds, which has indeed been shown to be the case. The evolution of spirit maturation in the cask can then be expected to be dependent upon fill strength, even if the cask and the conditions of maturation are identical. Some of the similarities and differences in spirit strength and maturation are indicated in Table 4.5.

The study of whisky maturation is hampered by a number of factors. Firstly, the time-scale over which maturation occurs is of the order of years, rather than weeks or months. Secondly,

Table 4.5 Typical capacity and maturation parameters for Scotch whisky production (data from Refs. 6 and 10).

Distillery	*Capacity/ mlpa[b]*	*Still strength/ %ABV*	*Fill strength/ %ABV*	*Casks*
Ardbeg	1.1	69.5–70.5	63.5	98% Ex-bourbon, 2% ex-sherry
Balvenie	5.6	>70	63.5	Ex-bourbon, -sherry and -port
Blair Athol	1.8	70	63.5	Refilled or first-fill ex-sherry, ex-bourbon
Edradour	0.1	69	63.5	Ex-Oloroso, ex-bourbon
Glenlivet	10.5	68	63.5	Ex-bourbon, ex-sherry, plain casks
Isle of Jura	2.2	70	63.5	Refilled American hogsheads, ASB, sherry butt
Knockando	1.3	70	63	Ex-bourbon, ex-sherry
Lagavulin	2.3	68.5	63.5	Third-fill ex-bourbon and third-fill ex-sherry
Laphroaig	2.9	67.5	63.5	First-fill ex-bourbon
North British[a]	68	94.5	62–68	Various
Royal Lochnagar	0.4	n/a	63.4	Ex-sherry puncheons and butts. No ex-bourbon

[a]Continuous, grain distillery.
[b]Million litres per annum.

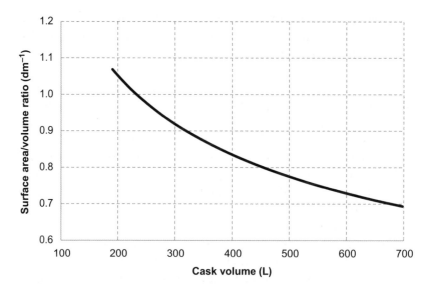

Figure 4.13 Dependence of the surface area/volume ratio of smooth casks on
cask volume. The reduced concentration of the internal cask surface in larger casks slows the maturation process. Experiments to
increase internal surface area (*e.g.*, by inserting additional wood
into the body of the cask) have demonstrable benefits, although
this practice is not universally permitted for the production of
whisky.

maturation is scale-dependent, as reducing cask size from around
190 L to a suitable laboratory scale will result in a substantial
distortion of the surface area/volume ratio (Figure 4.13), in turn
affecting the trajectory of maturation. Thirdly, there are variations
even between casks from the same wood that have similar histories,
indicating that there is a statistical requirement for repeat experiments. Thus, while experiments can be miniaturized and sped up
(the latter by increasing temperature), there is a risk that observations under such conditions will not reflect behaviour at the
commercial scale.

It is fair to say that the mechanisms of whisky maturation have
not been fully elucidated and indeed it is difficult to assess the
progress of maturation based on a range of analytical indicators.
The maturation of Scotch whisky has been increasingly studied as
analytical techniques have developed. Nonetheless, while maturation can be followed in terms of its detailed dynamic chemical
profile, it is still far from clear how this relates to the sensory

properties of the final product. Not only is there a non-linear relationship between concentrations of flavour compounds and the intensity of their perception, but also there are demonstrable interactions between flavours that can affect the overall flavour profile of the product. Such relationships, however, can only be demonstrated by the extensive and well-designed sensory experiments, as it is not possible to readily demonstrate such synergistic or antagonistic effects by conventional analytical means.

The chemical reactions that occur during whisky maturation (Figure 4.14) fall into two broad categories: additive and subtractive. Some authors refer to a third category, interactive, but here interactive specific reactions are considered to be either additive or subtractive depending on whether the main quality impact is removal of reactants or the formation of products.

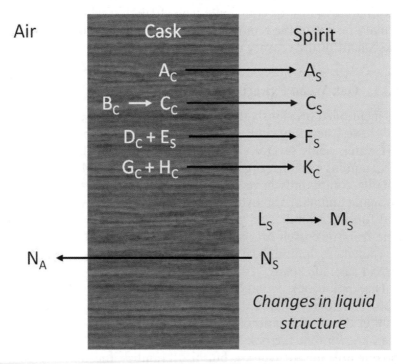

Figure 4.14 A summary of the mechanisms of whisky maturation. The subscripts refer to the localization of a given molecular species; *i.e.*, A = air, C = cask and S = spirit. 'Changes in liquid structure' refers to the formation of ethanol clusters as maturation proceeds.

The term additive is used to include reactions that introduce new compounds into the spirit or that elevate their concentration. A clear example of an additive reaction during whisky maturation is the extraction of chemical entities from the cask, be they from the raw heartwood, the toasted and charred layers of the cask or from the residual liquid and adsorbed components from whatever was used in the previous fill. These extractives can also be supplemented by the solvolysis of wood constituents during maturation and interactions between the wood and the spirit. Subtractive reactions result in the net loss of specific chemical components from the maturing spirit. This can occur by physical processes (such as evaporation out of the cask), adsorption or chemical degradation at the charred layer of the cask, and chemical degradation re-actions, such as oxidation. If the impacts on the spirit are con-sidered in sensory rather than chemical terms, then in this context additive reactions tend to enhance the perceived sensory impact of existing attributes, or contribute new attributes, whilst subtractive activity results in the removal of given sensory attributes, often to the benefit of the sensory quality of the product.

4.3.1   Oak Wood Extracts

As mentioned previously, both American white oak and European oak have ostensibly similar compositions. Extraction of oak with hot water yields several classes of compounds, including volatile oils, tannins, sugars, organic acids and sterols. More than 100 volatile compounds have been isolated from oak, which include aromatic and aliphatic hydrocarbons and acids, as well as phenols and furans. As indicated previously, compounds that are extract-able from oak itself are generally not replenished if a cask is re-juvenated, so limiting the scope of rejuvenation as a method to extend cask life-spans.

The major oak wood components are the so-called oak or whisky lactones (Figure 4.7). They occur in two forms, differing only in the configuration of the carbon at position-5. It has long been known that the oak lactones have been a contributor to the flavour not only of whisky, but also brandies and matured rums. As is common for flavour compounds, changes in stereochemistry can markedly affect their flavour perception. Recent work by Noguchi *et al.*[4] determined that the flavour threshold of the *cis*-oak lactone was some 5.5 times lower than its *trans*-counterpart (0.15 *vs.* 0.83 mg L^{-1} in 20% (*v/v*) aqueous ethanol). However, they also reported that there is a synergy between the two, with somewhat additive behaviour intensifying the perception of a 'coconut' flavour. This observation typifies the complexity of sensory experiments.

In common with other plant sources, the tannins from oak are chemically complex. They tend to be either the hydrolysable tannins, based on gallic acid (gallotannins) or ellagic acid (ellagitannins) or condensed tannins, based on the flavanol moiety (Figure 4.15). Tannins are extracted relatively rapidly into the maturing spirit, with the early and relatively rapid extraction rate slowing during the first year of maturation. Typical phenolic contents of Scotch whiskies (in gallic acid equivalents) are generally lower (<20 mg L^{-1}) than in American whiskies (up to 60 mg L^{-1}), presumably because there are greater levels of extractives in the

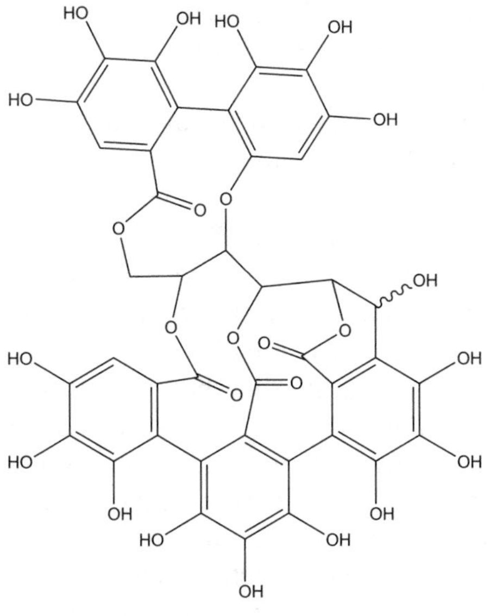

Figure 4.15   Optical isomers of vescalagin (aliphatic hydroxyl in R-configuration) and castalagin (aliphatic hydroxyl in S-configuration) as two examples of the ellagitannin structure. The extraction of compounds such as these from oak wood is thought to assist in the enhanced formation of dynamic ethanol clusters in matured whisky.

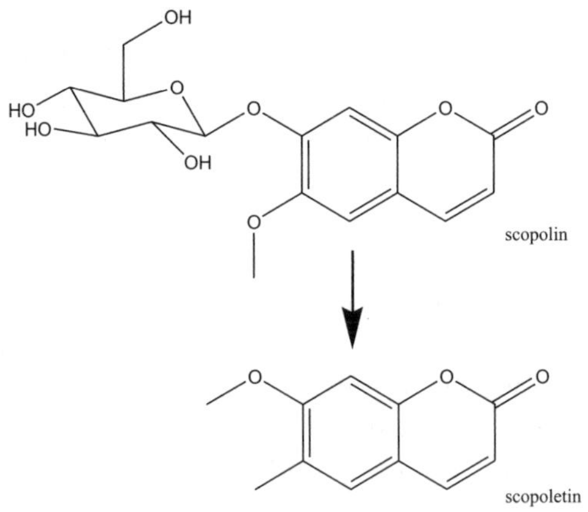

Figure 4.16 Proposed formation of scopoletin during the seasoning of oak wood.

new casks used in the latter case. Additionally, the fluorescent compound scopoletin, released from the glycoside scopolin during wood seasoning, (Figure 4.16) is also extracted into spirit during maturation. Scopoletin can be present in matured whisky at concentrations of up to 1.5 mg L^{-1} of pure alcohol.

Sugars and glycerol found in finished whisky are non-volatile and therefore come from the cask, their concentration increasing as spirit ages. In common with the tannins, their rate of extraction generally decreases over time, but they are not extracted as readily as tannins. The major sugars found in whisky are arabinose, glucose, xylose and fructose, while the total levels of sugars found in matured whisky can range from 10–25 g L^{-1} of pure alcohol.

A plethora of non-volatile dicarboxylic acids can be extracted with water from oak wood, but perhaps surprisingly are not found in matured spirit. Rather, these acids undergo esterification, presumably catalysed by the relatively acidic spirit, resulting in the formation of diethyl esters of fumaric, succinic and azelaic (nonanedioic) acid. Compounds, such as diethyl succinate, are considered to play a role in the mouthfeel and add 'legs' to the sensory performance of wine. However, their impact on the flavour of whisky is not clear at present. In all cases, the acids should be extractable into the spirit and their aqueous solubility reduced by the action of esterification.

Several sterols – predominantly β-sitosterol – have been identified in matured whisky. This is of relevance as they are poorly water-soluble and indeed they have been identified both as a component of flocculated material in whisky and in the residues from whisky filter sheets. Levels of such compounds are likely to be higher in matured spirits that are filled into cask at higher alcohol strengths as they will be more soluble at higher ethanol concentrations and, therefore, more readily extracted from the wood.

As mentioned previously, the most striking change to maturing new make spirit is the relatively rapid development of colour. This is not always apparent as spirit is matured in essentially closed casks, but experimental systems that partly contain spirit in glass (Figure 4.17) clearly show that colour is extracted rapidly early in the maturation period and again slows as maturation advances. Colour progression is typically lighter then darkens yellow before taking on amber and reddish-yellow hues. The colour of a spirit can be modified by the judicious choice of casks, so that maturation in Fino

Figure 4.17 Small-scale maturation system used at Heriot–Watt University. The glass dome allows the progress of maturation to be observed. Typically the base is charred American oak, with a surface area/ volume ratio of around 1 dm^2 L^{-1} (similar to that of an American standard barrel).

and Amontillado casks results in lighter colours, whilst Olorosos will give richer, darker colours. Colour assessment is also a useful guide to understanding the performance of casks during maturation. An exhausted cask will give a spirit with an insipid, pale yellow colour, even after years of maturation, as there is very little in the way of colour bodies remaining in the cask. Such a spirit is unlikely to have a role either as is or as part of a blend in a final commercial product and this is a situation that is to be avoided.

4.3.2   Changes in the Components of New-make Spirits

The compounds present in the new make spirit can undergo either physical loss, by evaporation, or chemical change. The losses of ethanol and water are significant over the course of whisky maturation. In the UK, losses amount to 1–2% of the volume per annum, whilst in India and Australia, evaporation rates can be as high as 12% per annum. In the UK, the losses of ethanol are greater than those of water, so that not only the volumes in the cask decrease, but also the strength of the maturing spirit decreases. If the strength drops too low, then it may be problematic to use this spirit either as is or in a blend. Clearly, filling at higher ethanol concentrations delays this problem to a later maturation date. In the USA, the opposite is the case, so that water is lost from the cask more readily than ethanol. Presumably, this is due to a combination of both higher temperatures and a lower relative humidity in the warehouse. Fill strengths in the USA are limited to no more than 62.5% ABV in recognition of this behaviour. Other compounds have also been shown to be lost in model systems by evaporation, including acetaldehyde, propan-1-ol, ethyl acetate and *iso*-amyl alcohol.

The source of acetaldehyde has been shown to be predominantly *via* ethanol oxidation. Furthermore, there is a clear effect of the charred wood itself, catalyzing the oxidation reaction. Presumably, the ethanol that permeates the wood structure is exposed to atmospheric oxygen in the proximity of the char to allow the reaction to take place. Mechanistically, this is most easily rationalised by the diffusion of molecular oxygen into the cask through stave joins. Similarly, acetic acid is also derived mainly by ethanol oxidation. However, this is not the full story. Radioactive labelling experiments have indicated that acetate is derived from another source, attributed to the acid hydrolysis of acetylated wood species. Similarly, caprylic (C_8), capric (C_{10}) and palmitic (C_{16}) acids have been shown to increase substantially during maturation experiments.

It is perhaps unsurprising, in light of the development of acetic acid, that ethyl acetate levels increase during maturation. Similarly, where the concentrations of other organic acids also increase, their esters are also likely to be enhanced. This is due to esters and water being in equilibrium with their acid and alcohol counterparts, so as acetic acid levels increase, the equilibrium shifts towards the ester. Although ethanol is lost during maturation, its concentration is so high that to a first approximation it can be considered to be constant in this context. The fate of other esters during maturation is less clear, generally remaining unchanged or decreasing over the course of maturation.

A key reaction during maturation is the loss of acrolein. As a pure compound it has both lachrymatory properties and is pungent. Research has indicated that acrolein reacts over a period of years with ethanol to form 1,1,3-triethoxypropane. Essentially, two reactions are occurring here: the 1,4-addition of ethanol to the acrolein carbon–carbon double bond and the formation of the diethyl acetal (Figure 4.18).

In terms of flavour impact, few classes of compound can be more significant than those containing sulphur (Table 4.6). Compounds are typically flavour-active at the μg L^{-1} level and, whilst their analysis has been aided enormously by the development of specific gas chromatographic detectors, they can prove problematic both to measure and to control. Compounds, such as DMS, are appreciably volatile (boiling point, b.p., 38 °C) and it would be reasonable to expect DMS losses to be attributed to evaporation. Isotopic labeling studies have shown that the major pathway for DMS loss is its oxidation to dimethyl sulfoxide (DMSO), catalysed by the char present on the inner surface of the cask. Furthermore, DMSO can undergo further oxidation to non-volatile dimethyl sulphone. Both oxidation products occur in malted barley but are insufficiently

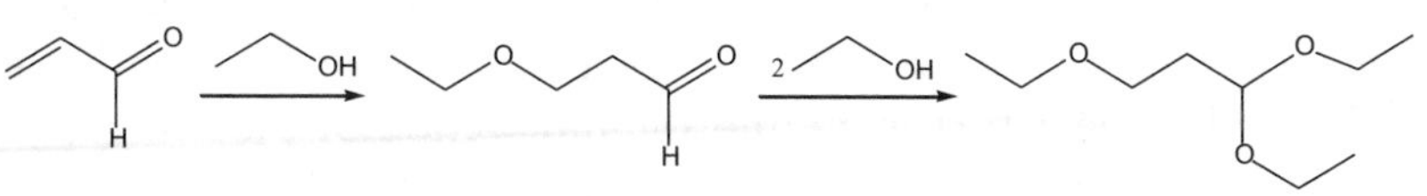

Figure 4.18 Proposed pathway for the loss of the pungent compound, acrolein, in maturing spirit.

Table 4.6 Some sulphur compounds in whisky, their flavour descriptors and fate during maturation (based on Ref. 5).

Compound	Descriptors	Fate during maturation
Dimethyl sulfide (DMS)	Tinned tomato, sweetcorn	Lost by evaporation and oxidation
Dimethyl disulfide (DMDS)	Rotten vegetable	Lost slowly
Dimethyl trisulfide (DMTS)	Onion	Essentially unchanged
Methional	Cooked/mashed potato	Quickly lost
3-(Methylthio)propanoate	Vegetable, onion, garlic	Lost within the first few years
3-(Methylthio)propyl acetate	Sulfurous	Lost within the first few years
5-Methyl-2-thiophene-carboxaldehyde	Fruity, benzaldehyde-like	Unchanged
Benzothiophene	Solvent, rubbery	Unchanged
Benzothiazole	Sulfurous, rubbery, coffee	Unchanged

volatile to be recovered in distilled spirit. Dimethyl disulphide (DMDS) is also lost slowly during maturation, but levels of the highly flavour-active dimethyl trisulphide (DMTS) seem to be unaffected by the maturation process. Other sulphur compounds, characterized by the presence of heterocyclic sulphur (such as thiazoles or thiophenes), are similarly not affected by maturation. Thus, the presence of these latter flavour-active compounds in new make spirits going into the cask will primarily determine the sulphur flavour characteristics in the final matured spirit.

4.3.3 Changes in the Wood-derived Lignin Compounds

The lignin-derived compounds are of substantial importance to the development of what is understood to be matured whisky. Initial work on brandy identified key components, such as vanillin, syringaldehyde, coniferaldehyde and *p*-hydroxybenzaldehyde (Figure 4.12), the presence of which was confirmed shortly afterwards in malt whisky and other matured spirits. The pathways for the formation of lignin derivatives in Scotch whisky and other oak-matured spirits (Figure 4.19) are clearly influenced by both the wood and its heat treatment and the interaction of the spirit, predominantly ethanol, on the extracted components. The importance of heat treatment is evident from experimental observations of lignin-derived compounds, increasing by two or three orders of

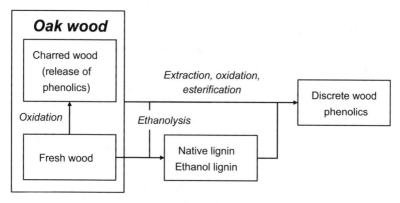

Figure 4.19 A simplified schematic outlining lignin degradation pathways. Discrete analytes may be extracted directly from the fresh or charred wood, whilst the ethanol-rich spirit has been shown to extract both native and soluble or ethanol lignin. These can undergo further breakdown to yield discrete molecular entities. Lignin-derived compounds can also undergo oxidation at the cask/spirit boundary, a mechanism considered to be of importance in the Cognac and whisky contexts. (After Ref. 9.)

magnitude in charred relative to uncharred oak. Research has also indicated that some oak lignin is sensitive to oxidation when in contact with spirit, such that the concentration of lignin derivatives is an order of magnitude higher on the inner surface of staves compared to the outer surface.

4.3.4 Changes in Spirit Properties during Maturation

Without exception, the relative concentration of ethanol and water will change as maturation proceeds. This is important as the following properties of ethanol–water mixtures will change according to their relative concentrations:

- Density – decreases non-linearly but monotonically as ethanol content increases
- Surface tension – decreases with increasing ethanol concentration
- Viscosity – reaches a maximum at close to 40% ABV, of around three times that of water
- Polarity – decreases monotonically as ethanol content increases

The combination of surface tension, viscosity and solvating power will influence the ability of the spirit to penetrate the wood, extract wood components and the rate of diffusion of substances into and out of the wood.

The mixing of water and ethanol also results in heat evolution, suggesting changes in the molecular structure of the mixture. There is also an associated volume contraction. It has been shown that, above a certain, critical, ethanol concentration of around 17% ABV in water, the mixture is no longer homogenous and undergoes a phase separation into dynamic ethanol-rich clusters distributed in a relatively water-rich continuous phase. Furthermore, it appears that both fermentation volatiles and wood extractives promote the formation and stabilisation of ethanol-rich clusters. From a whisky-maturation perspective, the ethanol concentration is always well above the critical concentration for cluster formation, so that from filling to disgorging of the cask, the heterogeneous structure of the spirit persists.

4.3.5   Maturation and Spirit Strength

As we have already mentioned, the fill strength of new make spirit into casks is elective to some extent, typically ranging from 63–71% ABV. The impact of changes in spirit strength on extractive yields is apparent from reported research (Figure 4.20). However, in the case of more polar extractives, such as glycerol and monosaccharides, higher spirit strengths result in lower rates of extraction. The causes seem to be two-fold, with both rates of hydrolysis of, in particular, hemicelluloses and their solubility being enhanced under conditions of higher water activity. Based on these observations, it is perhaps unsurprising that a commonly accepted fill strength for Scotch whisky maturation is 63.5% ABV.

4.4   WAREHOUSING OF WHISKIES DURING MATURATION

The storage of casks to allow maturation to proceed is of paramount importance to the final quality of the matured spirit. Casks can be stored upright (*i.e.*, palletized), up to six casks high, or on their sides in a rack up to 12 casks high. The requirements for warehousing are two-fold: to ensure that the spirit is sufficiently

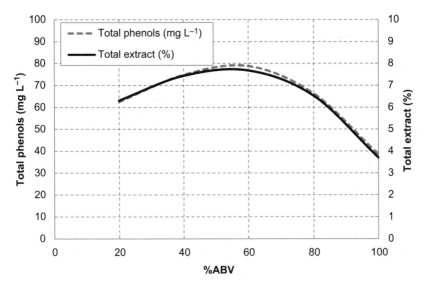

Figure 4.20 Impact of spirit strength on total extract and phenolics extracted from oak chips after five months of storage on ethanol–water mixtures. Clearly, there is a strong correlation between total extract and total phenolics. The latter were determined in gallic acid equivalents.
(Based on data from Ref. 7)

secure, not least because duty has not been paid at this stage, and to help manage the ambient conditions within reasonable limits of ventilation, humidity and temperature. Higher maturation temperatures generally give rise to faster rates of evaporation of both water and ethanol from casks, whilst humidity will affect the relative rates of ethanol and water loss. Thus, in warehouses of high humidity, ethanol losses are accentuated relative to those of water. Indeed, there are temperature and humidity gradients in a given warehouse, so that the maturation of ostensibly similar casks at different locations within the same warehouse can show observable differences in their time-dependent composition. In terms of overall losses from casks, work carried out on casks stored under controlled climatic conditions has neatly demonstrated the interplay between relative humidity and storage temperature on evaporative losses (Figure 4.21).

In other parts of the world it has sometimes proved necessary to mature under climate-controlled conditions, but in Scotland this is

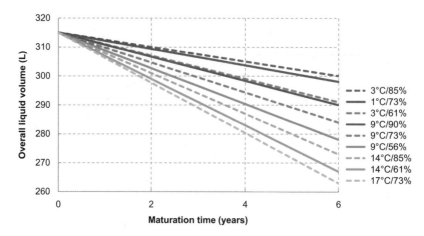

Figure 4.21   Influence of maturation and relative humidity on evaporative
losses from oak casks.
(Based on data from Ref. 8.)

not considered necessary. The variations of climatic conditions
across Scotland are relatively small, with reports of mean temperatures ranging from 7.4–9.7 °C, daily fluctuations ranging from
9–20 °C and relative humidity ranging from 70–80% throughout
the year. Thus, in terms of the relative importance of the principal
factors affecting Scotch whisky maturation, they can be ordered
as follows: *cask* ≈ *distillery* > *warehouse location*. Finally, it is worth
remembering that while ambient temperatures may fluctuate on
a daily basis, the insulating properties of the oak cask mean that,
in practice, the spirit temperature will only fluctuate modestly.
Significant seasonal temperature swings, however, are more likely
to be translated to noticeable changes in cask spirit temperature.

4.5 QUALITY ASSURANCE AND ASSESSMENT
OF MATURED SPIRIT

Given the amount of time it takes to produce fully matured spirit, it
is essential to implement all measures that are feasible to minimise
the risk of unacceptable performance during the maturation process. Generally, the cask is not broached throughout maturation
so, until the bung is removed, there is little way of telling whether

the spirit has developed satisfactorily. The three factors that can be managed prior to the onset of maturation are spirit composition, wood provenance and chemistry, and the physical attributes of the cask. The former is relatively straight-forward for the experienced distiller and aspects have been discussed in the previous chapter. The provenance of wood is also feasibly managed, although the logistics of managing cask supply is not trivial. Thus, with an estimated production in 2012 of around 445 million litres of pure alcohol pa in new make spirit, assuming an average cask volume of 200 L and an average fill strength of 67% ABV, around 3.3 million new casks are required to store the spirit. With stocks running close to eight years, then a reasonable estimate of casks to manage just Scotch whisky is around 26.7 million! Nonetheless, software solutions allow effective tracking of casks so that their provenance is largely assured. The quality of the cask surface in the context of whisky maturation, however, is not only dependent on the liquid that was previously kept in the cask, but also on the cask storage conditions between fills. An odour check of any cask prior to filling can reduce the incidence of potentially damaging development of musty taints and off-flavours due to fungal infections in previously empty casks. It is essential of course that the physical integrity of the cask is such that there are no unreasonably large losses from the cask due to leakages. This is more of an economical consideration, although more rapid losses can, in principle, lead to more rapid ingress of oxygen into the cask headspace.

For spirit leaving the cask there are several clues to the success, or otherwise, of the maturation process. Visual inspection or spectrophotometric measurements of the spirit colour can give a clear indication of the level of extractives recovered from the cask. Odour assessment can again be used to ensure that there are no musty taints present, which would make it difficult to use the spirit even in a blend. Nosing of the spirit in terms of its balance and desirable aroma properties can help to reassure the experienced blender that the spirit is as expected. This is important not only for the bulk of spirit that will end up in blends but also for single malts. This is because even a single malt is a combination of the contents of a number of casks so there is a need to ensure that no atypical flavours present themselves in the final product. Such assessment is even more critical for those producers that package offerings from single casks.

4.6   IN CONCLUSION...

The qualities of whisky are substantially reliant on the process of maturation in oak casks. This is a cost-intensive process, both in terms of the cost of procurement and the management of cask inventory. Together with the cost of warehousing for the protracted maturation period and the time to recover investment costs, it is evident that the production costs for whisky can be expected to be significantly higher than for unaged spirits. The benefits of maturation, however, are clear: changes in the volatile compounds' composition gives rise to a spirit of improved sensory properties, whilst the non-volatile extractives contribute to the flavour and the structure of the spirit at the molecular level. The progress of maturation is dependent on a number of factors, including the composition of the spirit (in particular its ethanol content), the physical integrity of the cask, its provenance and the time and conditions under which maturation occurs.

NOTES AND REFERENCES

1. F. P. Hankerson, *The Cooperage Handbook*, Chemical Publishing, Brooklyn, 1947.
2. So-called because saturated aqueous ammonia contains 880 mL ammonia per mL water.
3. See The Scotch Whisky Association, http://www.scotch-whisky.org.uk/ (accessed 30th July 2013).
4. Y. Noguchi, P. S. Hughes, F. G. Priest, J. M. Conner and F. Jack, *Proceedings of the 2008 Worldwide Distilled Spirits Conference*, eds. G. M. Walker and P. S. Hughes, Nottingham University Press, Nottingham, 2010, pp. 243–251.
5. M. Masuda and K. Nishimura, *Journal of Food Sciences*, 1982, **47**, 101–105.
6. A. S. Gray, *The Scotch Whisky Industry Review*, 35th edn, Sutherlands, 2012.
7. L. Nykanen, Aroma compounds liberated from oak chips and wooden casks by alcohol. In: *Proceedings of the Alko symposium on flavour research of alcoholic Beverages*, eds. L. Nykanen and P. Lentonen, Helsinki, Finland, June 1984, Foundation for Biotechnical and Industrial Fermentation Research. 1984, pp. 141–148.

8. J. M. Philp, Cask quality and warehouse conditions. In: *The Science and Technology of Whiskies*, eds. J. R. Piggott, R. Sharp and R. E. B. Duncan, Longman. Harlow, Essex. UK. 1989, pp. 264–294.
9. K. Nishimura, M. Ohnishi, M. Masuda, K. Koga and R. Matsuyama, Reactions of wood components during maturation,. In: *Flavour of Distilled Beverages: Origin and Development*, ed. J. R. Piggott, Ellis Horwood, 1983, pp. 241–255.
10. M. Udo, *The Scottish Whisky Distilleries*, Black and White Publishing, Edinburgh, 2006.
11. P. S. Hughes and J. A. Hughes, *Proceedings of the 2011 Worldwide Distilled Spirits Conference*, eds. G. M. Walker, I. Goodall, R. Fotheringham and D. Murray, Nottingham University Press, Nottingham, 2012, pp. 87–93.

CHAPTER 5

From Blend-to-Bottle

5.1 INTRODUCTION

As discussed in previous chapters, the spirits that are selected
for bringing together for the final product have qualities that are
a function of the new make spirit, the cask history and the environmental conditions of maturation. There will be variations in
colour and flavour attributes of the final spirit from each cask and,
thus, to maintain consistent quality even of single malts, it is essential to bring together the contents of a number of casks to even
out the sensory idiosyncrasies of any given cask. (Only for single
cask bottlings, where conformance to a brand specification is arguably less important, can this be avoided in principle, although of
course it would be unwise to package a product with clear defects.)
Whisky brands that avoid the use of type I caramel for colour
adjustment have less room to manoeuvre when casks are being
combined, but managing colour quality is more straight-forward in
that it can rely more heavily on instrumental analysis. For the
production of blended whiskies, the sale volumes of which outstrip
those of malts by an order of magnitude, the combination of spirits
can require anywhere from 20–50 malts and 2–5 grain whiskies.
The result is a whisky that tends not to exhibit overly dominant
flavour attributes but, rather, will generally have a consistent and
rounded flavour.[1]

The Science and Commerce of Whisky
By Ian Buxton and Paul S. Hughes
© Buxton & Hughes, 2014
Published by the Royal Society of Chemistry, www.rsc.org

Indeed, the milder sensory attributes of blends is thought to be one of the reasons for whisky becoming as popular as it is today. Up to the third quarter of the 19th century, gin and brandy were the distilled drinks of choice in the UK. However, taxation on gin became increasingly punitive and brandy production, which requires the distillation of fresh wine, suffered firstly from a mildew infection of French grapes, then the outbreak of an American aphid known as phylloxera. These successive disasters left a gap in the spirits market, and the innovation of the whisky blend, attributed to Andrew Usher in the early 1860s, was ripe for exploitation by the so-called whisky barons towards the end of the 19th century. They successfully marketed and sold blends to the English market, recognizing that the English drinker preferred the taste of blends over the more heavily flavoured single malts. By the time brandy volumes began to recover, Scotch whisky already had a firm foothold, particularly in the UK and American markets.

5.2 CREATING THE BLEND

The blends available to the consumer today are carefully constructed by blenders with unique, recognizable and consistent flavour and colour attributes. The role of the master blender is still a vital one in many whisky companies, but today the scale of the task (for which more than 14 000 000 nine-litre cases of blended whisky were produced in Scotland in 2011)[2] is such that the blending operation in larger companies relies on is performed by a team. The construction of blends is very heavily reliant on expertise in flavour perception and this, together with the size and value of the blended whisky market, means that the composition of blends is a carefully guarded secret.

Unlike conventional instrumental analysis, flavour attributes are generally non-additive, so *a priori* prediction of flavour attributes based on analysis is fraught with difficulty. For this reason, whisky blending is often thought to be more of an art learned than a science and indeed blending is less a sequential task of bringing together spirits one after the other but rather an iterative one as the flavour attributes of the brand are built up. Certainly, whilst a blend may have a recipe involving specific whiskies at various proportions, these will need to be varied in recognition of the variation between whiskies on a batch-to-batch and a cask-to-cask basis. Also, it is not uncommon for specific whiskies to become unavailable either because a spirit becomes restricted or a distillery closes, so the blender needs to be able to make the appropriate substitutions in order to maintain the continuity of the final product. It is clear that producing a successful and consistent blend requires substantial expertise from the blender coupled with innate talent for sensory evaluation.

One way to consider the construction of a blend is to view the grain whisky components as the foundation of the blend, to which malts with both higher levels of fermentation and, often, maturation congeners, as well a range of distinct sensory attributes (*e.g.*, peaty, woody, sulphurous) are added to create the desired flavour profile. This is simplistic as grain whiskies, while generally lower in flavour intensity than their malt counterparts, do indeed vary in their flavour attributes, and play a key role in softening or rounding out more strident flavours in malt components. Because of the variation between grain whiskies, it is rare for a blend to rely on one grain whisky, and up to five may be employed.

The malt components can contribute the whole range of malt whisky flavour attributes. The large number of malts available, not only in terms of distillery but different malts (*e.g.*, length of maturation, cask provenance), can mean that the blender is almost spoilt for choice. However, it is important that malts are used judiciously to create the required style and that, in the event of a supply difficulty, there are options for substitutions or adjusting the recipe to ensure consistency. The more malts that are used, the more the risk of inconsistency is mitigated, hence why it is not uncommon for a blend to contain 20–50 malts. Heavily flavoured components, such as the Islay peated or smoky malts, are likely to be used in small quantities or they will overly dominate a blend resulting in an unbalanced product. In contrast, lowland malts are generally less distinctive in terms of the palette of single malts and can be used in larger quantities to add weight without too much character to a blend. Indeed they are often essential for the production of a good blend. Malts from the Highlands are numerous and diverse but generally have excellent reputations as single malts and as blend components, with distinctive sensory attributes that, with the Islay malts, are often the major flavours in final blends (Figure 5.1).

It is erroneous to consider grain whiskies as cheap substitutes in a blend, even if, as a commodity, grain spirit is of lower cost than malt

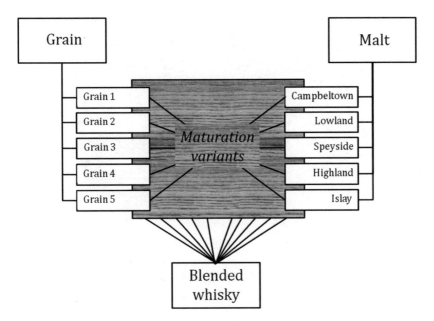

Figure 5.1 A schematic of Scotch whisky blending. A number of malt and grain new make spirits are matured in a variety of casks for different periods of time and brought together to create the final blend.

spirit. The proportion of grain whisky in a blend is not necessarily an indication of the blend quality. This is dictated by both the quality of the component whiskies and the skill used in synthesizing them into a final blend. However, there comes a point where over-reliance on grain whisky will be unlikely to result in a good blend.

Before starting the blending process, the blender needs to assess the component spirits. This is often carried out by sniffing or 'nosing' the spirit, although there may be a need to taste some of the potential blend components also. The samples are drawn off into nosing glasses (Figure 5.2) and assessed as is or, more often, diluted to around 20% ABV to open up the flavour-yielding volatile compounds and reduce the pungency of the alcohol present. Decisions on the sensory performance of the spirits are required. Firstly, are off-notes present? Secondly, is it typical of the style expected for that component? And, thirdly, if so, is it of the right overall quality for inclusion in the blend?

The practical management of blend production is obviously dependent on the availability of its various components. If these

Figure 5.2   Typical 150 mL tulip nosing glasses. The shape facilitates the swirling of the spirit whilst retaining spirit volatiles in the body of the glass. Left: new make spirit that has failed the misting test; right: matured whisky.

are from within the same distilling company, then access to forecast data should provide an early warning of stock depletion and trigger consideration of a shift away from an increasingly scarce resource. Forecasting a market three, five or even ten years ahead is an extremely challenging task and, if not performed sufficiently accurately, could have adverse implications certainly for bigger volume brands that rely on correspondingly bigger volumes of the various components. In Scotland, it is common to purchase or exchange stock with competitors, which exacerbates the forecasting difficulty. Nonetheless, it is worth pointing out that, in principle, with spirit from around 100 malt distilleries and seven grain distilleries, which are often further differentiated by the time spent in the cask and the past history of the cask, Scottish distillers have an unrivalled palette of whiskies at their disposal to maintain current blends and develop new ones. This in itself creates a substantial barrier-to-entry for distillers operating outside Scotland.

What has been said about blending up to now refers to the production of Scotch whisky. The production of blends in other parts of the world has some similarities to those activities practiced in Scotland. Japanese blends rely on the use of both grain and malt whiskies and, indeed, grain whiskies are a legal requirement. Cross-trading between Japanese producers is not known and given the relatively low number of distilleries in Japan, it is not surprising that Japanese distillers have worked assiduously to produce different malt whiskies at each plant. Additionally, whiskies have been imported in bulk into Japan in the past to broaden the range of available blending components. However, whilst shipments of malt and grain whiskies reached a high of nearly 15 000 000 L of pure alcohol in 2001, by 2007 this had dropped to a mere 10 000 L of pure alcohol, presumably due to overstocking.

The Irish distilling situation is quite different to that in Scotland. There are relatively few whisky types available for blending and, consequently, blends have a wide range of grain/pot still whisky ratios, ranging from light- to full-bodied. As discussed in chapter 3, Irish whiskies are made with a number of other small grains and this, together with blending of spirits of different ages and even using new oak casks, all provide additional variation that helps Irish blenders to expand their product map.

In some countries a greater range of constituents can be used to create blended products. For instance, neutral spirit is permitted in India and flavourings, including grape-based products (*i.e.*, wine, sherry, brandy) and rums, can be used within the local definitions of whisky. Different views as to what constitutes whisky had led to prolonged debate and erection of trade barriers between, for instance, India and the EU. Indeed recent legislation in India forbids the use of the word Scotch on any Indian-made product, although Scottish imagery is still permitted. In Canada, whilst blending may be carried out in a similar way to that in Scotland, there is also the practice of pre-blending spirits before maturation and the addition of flavourings prior to packaging.

To create the blend, the component casks must be identified and co-located. It is essential that all of the required casks can be found and brought out of the warehouse, as the absence of even one or two key components could have an adverse effect on the final expression of the blend. Once the assembled casks have been passed fit by the blender, they are rotated such that the open bung port is inverted over a stainless steel trough, from where the spirit runs off (*i.e.*, disgorged; Figure 5.3) into a blending vat through a coarse

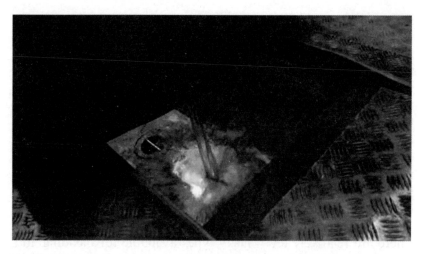

Figure 5.3   Whisky disgorging. The spirit is poured directly into a blending vat *via* a stainless steel trough on the floor.

mesh, typically around 5 mm. This is to trap the largest char fragments that are inevitably dislodged from the inside of the cask during disgorging. There is little need to be concerned about microbial infection at this point, as the strength of the spirit leaving the cask is usually significantly higher than 50% ABV. All of the spirits can be vatted together; although, for some distillers it has been the practice to vat grain and malt whiskies separately, bringing the two components together for bottling. Whilst the blender may be confident that the vatted product is acceptable, the blender nonetheless checks it to ensure it meets expectations. If not, there is still time to make some minor adjustments prior to bottling.

Some producers will follow the vatting stage with a so-called marrying stage, transferring the blended whisky into butts or larger wooden containers, where they are held for three to six months. There is no significant further maturation, as the marrying vats have long been exhausted of extractives. The marrying stage is optional, with blenders having different opinions as to the need for this additional step. Again, such decisions are based on the perceived sensory effects rather than hard analytical data. Once the blend has been finally approved, the spirit is ready for the final processing stages in preparation for bottling. From the blending vat, the spirit is once again strained, through a mesh of around 1 mm into the reducing vat.

5.3 FINISHING AND BOTTLING

To complete whisky production, water and spirit caramel (where required) is judiciously added at this point to adjust the spirit to sales strength but this, at least for Scotch whisky, must not be below 40% ABV, a clear stipulation of the 2009 Scotch Whisky Regulations. Most whiskies are then chilled and filtered, filtration typically being carried out through a plate-and-frame press with 5 μm filter sheets (Figure 5.4). The role of chilling is that it effectively forces spirit components that are sparingly soluble out of solution; the precipitates formed are then filtered whilst low temperatures are maintained. This reduces the risk that these sparingly soluble components will eventually precipitate in the bottle. The main components that precipitate are lipids extracted from oak during maturation and longer chain fatty acid esters. Dilution of spirit to sales strength already reduces the solubility of such lipophilic materials.

The decision to chill or not before filtration is one made by the brand team, with some producers concerned that chill filtration reduces some sensory aspects of the product, such as its body, mouthfeel and peaty character. The severity of the chill filtration operation is driven mainly by the degree of cooling applied, as deeper cooling will force more material out of solution. Typically,

Figure 5.4 Detail of a typical plate and frame filter press, showing the filter pads in position. The presence of condensation reflects the fact that the spirit being filtered has been cooled prior to filtration.

chill filtration is carried out at temperatures of -10 to $4\,°C$; although, $0\,°C$ can be considered typical for single malts, and perhaps slightly lower temperatures for blends. The filtered spirit is then passed to the bright spirit vat, from where it is drawn to the bottle filler, *via* a 10 µm guard filter. The consumer might spot whiskies that are not chill-filtered should they take whisky with water or ice, as these can induce the formation of visible haze. Whilst such haze formation might be visually apparent, the flavour impact will be minimal.

It is difficult to underestimate the impact that packaging has on both the integrity of the product and the marketing associated with branding. From a technological perspective, whisky tends to be packaged either in glass or in plastics, such as polyethylene tere-phthalate (PET). The latter has become increasingly popular both for lower cost products and in airports, for half bottles, as well as miniatures for on-board consumption. Bottles are typically cleaned prior to use, which is usually performed by an air jet to dislodge any particulate materials. Washing may also be performed but this is generally not required as most whisky bottles are non-returnable.

Whisky bottle closures are most commonly the so-called **ROPP** (roll-on pilfer-proof) caps and, for higher end products, cork-based closures. The former are used extremely widely for glass and plastic bottled beverages, including mineral waters, carbonated soft drinks, spirits and liqueurs. They have some key features: sealing the product within the package and showing clear evidence of pre-opening. Cork closures can convey a premium image, but cork management and selection must be carried out assiduously to minimise the risk of product spoilage, primarily through potential release of cork taint into the final product.

The labelling of the final packaged product, together with the geometry of the bottle itself, communicates the identity of the brand. In principle, labelling can be directly printed or etched on to the bottle, glued on or applied as a pressure sensitive label. In the UK, for all packs larger than 350 mL and more than 30% ABV, a duty paid stamp must also be visible for the product to be legally sold.

5.4 QUALITY MANAGEMENT

The quality of the final product will be dependent on the raw materials and process stages of production. By quality, here, we are considering those aspects of the product that are of relevance to the consumer, which is the visual (colour, clarity) and the flavour attributes of the spirit. We will limit quality considerations to the properties of the liquid product itself; although expectations of the brand prior to purchasing and consumption are particularly important to the Scotch whisky industry, given the iconic nature of many of the well-known brands. For the quality of whisky liquid, aspects of raw materials, fermentable extract, fermentation, distillation, maturation, blending/mixing and finishing are all considered.

A central tenet of the Scotch Whisky Regulations of 2009 is that no new make spirit can be prepared at more than 94.8% ABV, on the premise that further rectification would result in a spirit that did not retain the flavour attributes of its raw materials. The specifications set for malted barley are dependent upon whether it is intended for single malt spirit or for grain spirit, but are primarily for the purpose of process efficiency and consistency rather than the ultimate flavour qualities of the final product.

For the fermentation to perform as expected, the fermentable extract should be consistent. Primarily this is dependent upon both the concentrations and profiles of both fermentable sugars and of free amino nitrogen. Of course satisfactory fermentation performance also requires a range of micronutrients. Unlike the brewing context, however, the retention of amylolytic activity in the fermentation means that not only does the extract need to conform going into the washback, but ideally the enzymatic activity during fermentation should be consistent. To ensure the consistency of fermentation, the use of a consistent quality of yeast (*e.g.*, monitored by attention to viability, vitality, pitching rate) is essential for what is arguably the pre-eminent processing aid in the distilled spirits industry.

The distillation operation can substantially affect the composition of the new make spirit. Checks on wash strength, the cut-point from foreshots to spirit, the cut-point to feints, the %ABV of the foreshots/feints/low wines receiver and the rate of spirit production will all influence spirit quality. Arguably though, maturation is one of the highest risk areas of whisky production. Variations in casks and large variations in ambient storage conditions will influence final spirit quality, although the latter only tends to be significant in the event of damage to the fabric of the building.

5.5 INSTRUMENTAL ANALYSIS

To ensure that the final product is within specification, several routine and semi-routine tests can be employed. The most important analysis is that of alcohol concentration, which can be used to estimate the alcohol yield from raw materials (the predicted spirit yield, as discussed in chapter 3), ensure the correct alcohol concentration going into cask and to confirm package strength for HM Revenue & Customs (HMRC) compliance. Alcohol measurements can also be used to decide on cut points for changing to spirit and feints collection. Measurements can be either 'real' or 'apparent'. Real alcoholic strength is measured by distillation and gives a true result (although it is time- and labour-intensive), whilst apparent alcoholic strength is determined without distillation and is mainly based on density measurements. As ethanol is less dense than water, higher ethanol concentrations result in lower density readings. The term 'apparent' is used because the true result can be obscured by dissolved solids, which effectively increase the density and, therefore, give falsely low alcohol readings. This reduced reading is known as obscuration:

$$\text{Obscuration} = \text{Real} - \text{Apparent} \ (\% \ \text{ABV}) \tag{5.1}$$

The level of dissolved solids in matured whisky is generally low, with obscuration levels rarely exceeding 0.2% ABV. Traditionally, density measurements were determined by hydrometer, although, today, electronic density meters are favoured. In both cases, accuracy is typically ±0.1% ABV. Other methods include the use of appropriately calibrated near infra-red (NIR) spectroscopy, which has the advantage that it can give good performance for highly obscured products (such as liqueurs and lower strength ready-to-drinks). Gas chromatography (GC) is also useful but is not considered sufficiently accurate for tax purposes.

Another important analysis is that of the so-called major volatile congeners, principally *n*-butanol, isobutanol and 2- and 3-methylbutanols (Figure 5.5). The term 'major' implies the occurrence of these compounds at relatively high levels, often in excess of 5 g L^{-1} pure alcohol. These are most readily analysed by gas chromatography with a flame ionization detector (FID). Conveniently, this methodology also separates out acetaldehyde, methanol and ethyl acetate. Their formation is primarily during

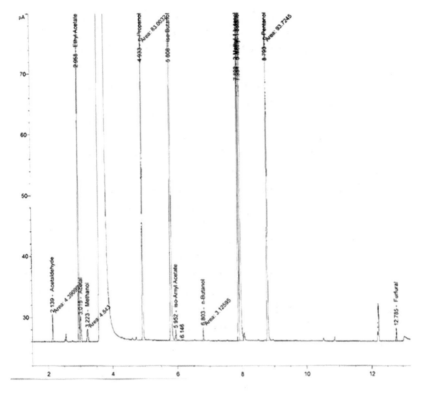

Figure 5.5 GC-FID analysis of major congeners from a bottled malt whisky (courtesy of the Scotch Whisky Research Institute).

fermentation but, as discussed previously, their levels can be affected by maturation and blending. The relative concentrations of the major congeners can indicate the source of the spirit. For instance, for malt spirit, the ratio of the C5 alcohols to isobutanol is usually in excess of 2.5, but for grain spirit, with the additional rectification afforded by column distillation, this ratio is well below 0.5. Blends and Canadian whiskies often have low or intermediate ratios here, indicative of the blending of malt and grain spirits. The major congeners can also be used to monitor the efficiency of grain spirit (and grain neutral spirit) production.

Occurring at lower concentrations (at or below mg L^{-1} levels), the trace volatile congeners are made up of longer chain alcohols, aldehydes and esters (apart from ethyl acetate). In this instance, some form of sample pre-concentration is required and a typical extraction of a 10 mL of sample with 1 mL of n-pentane (boiling point: 36 °C) immediately gives a potential 10-fold increase in concentration. This is then subject to GC analysis. The use of solid phase micro-extraction (SPME) is proving increasingly popular. Here, a thin hollow tube of absorptive material, or 'fibre', within a syringe needle, is introduced into the headspace or the body of the liquid and, with the appropriate selection of fibre, substantial concentration of the minor congeners is achieved. This fibre is then inserted into the injector of a GC and the heat in the injector thermally desorbs the sample. The process is easily automated and there is no requirement for solvent extraction. FID-based detection of eluted bands is common, although coupling to a mass spectrometer (MS) can both aid selectivity (*via* selected ion monitoring) and the identification of unknown peaks.

We have mentioned previously that sulphur volatiles are of substantial importance to the flavour of whiskies (Figure 5.6) and, because they are flavour-active at such low concentrations, analytical methods require a significant element of selectivity. Sulphur-specific detectors, such as the pulsed flame photometric detector (PFPD) and the Sievers chemiluminescence detector, are the most commonly used.

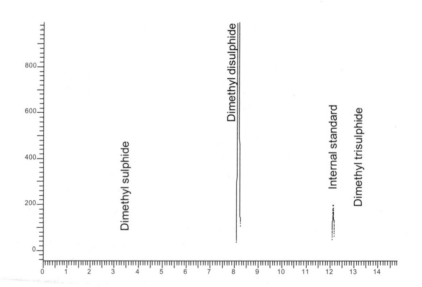

Figure 5.6   GC-PFPD analysis of sulphur volatile compounds from new make malt spirit (courtesy of the Scotch Whisky Research Institute).

The presence of non-volatiles in whisky essentially precludes the use of gas chromatography unless there are options for derivatisation that increase the volatility of the target analytes. The concentration of whisky solids is usually low, less than 2 g L^{-1}. During maturation, the spirit will acquire virtually all of the non-volatiles that are present in the final product, with ash (calcium, magnesium, sodium, potassium salts) being of the order of 0.2 g L^{-1}. Metal cations can be determined by ion chromatography or atomic absorption spectrometry. Residual sugars, extracted from the cask or from any spirit caramel added, typically occur at concentrations of $<200 \text{ mg L}^{-1}$. They can be conveniently measured by ion or ion exchange chromatography. Other detectors, such as the universal elastic light scattering detector (ELSD), are useful in that sugars have few chemical features that lend themselves to sensitive detection. High performance liquid chromatography (HPLC) or its more recent variant, ultra performance liquid chromatography (UPLC), are often the techniques of choice for the determination of non-volatile components. Typically, the columns used are reverse-phase and detection is often with ultra-violet (UV) or fluorescent detection. Compounds derived from wood, such as aromatic lignin-derived products, are UV-active and are readily detected. Some compounds, such as scopoletin, can be quantified in a highly selective way due to their fluorescence (Figure 5.7).

The pH of whisky at 40% ABV is around 4.0–4.5, although it can be higher if it is reduced with softened rather than demineralised water. However, with water concentrations typically of the order of 30 M (compared with 55.5 M in pure water), proton activity will be significantly lower, so it is important to calibrate pH probes at 40% ABV.

Arguably, one of the most recognisable characteristics of whiskies is the presence, in some cases, of a pronounced phenolic character. These are predominantly introduced during the production of peated malt (chapter 3) and, given their impact on the flavour attributes of peated spirit, there can be occasion to determine their levels in new make and in final products (Figure 5.8). To a first approximation, the relative proportions of the main phenolics are similar across peated whisky brands and, whilst it is not currently possible to produce a synthetic peaty flavour from combining known phenolics, major flavour impact compounds include the methyl-phenols (*i.e.*, cresols) and the methoxyphenols (*i.e.*, guaiacols).

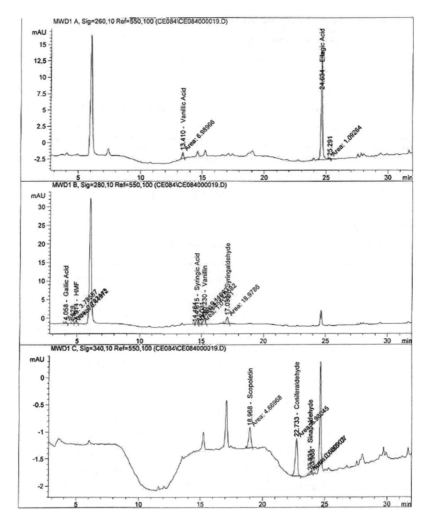

Figure 5.7 HPLC analysis of wood extractives found in a five year old malt whisky matured in American oak (courtesy of the Scotch Whisky Research Institute).

Before consumption, a consumer will almost always see the product, either in pack or in the serving glass. Visual aspects, such as haze and colour, are rapidly appraised and, if the consumer is familiar with the product, it will be subconsciously compared with what might be expected. Commonly whisky 'tint' is determined by measuring its absorption of visible light at around 500 nm.

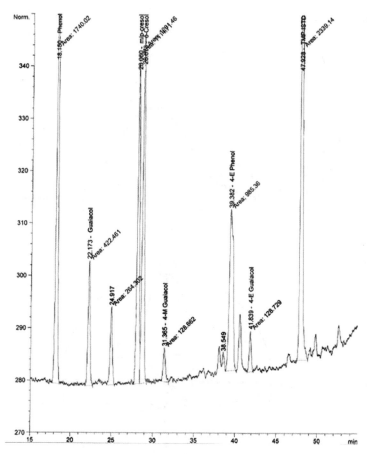

Figure 5.8 HPLC analysis of phenols found in a peated new make malt spirit (courtesy of the Scotch Whisky Research Institute).

Multiplying this absorbance by 100 gives the final tint value. However, the human eye perceives light over a range of wavelengths and the tint can give a good indication of the product colour, especially when considering the management of an established brand. However, when more detailed colour information is required, other more comprehensive measurements can be employed. For instance, Tristimulus measurements, based on the wavelength-dependent sensitivity to the various colour-sensing cones in the human eye (Figure 5.9a) can be used to map measured colours on to a three-dimensional colour space

(a)

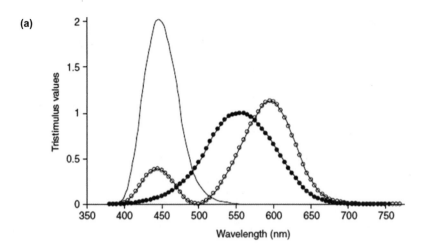

(b)

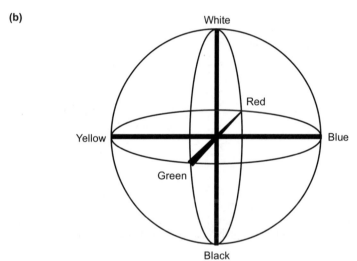

Figure 5.9 a) Each of the three colour-sensing rods in the human eye has a characteristic response to visible light. Tristimulus colour measurements allow products to be mapped on to the CIELAB colour space (b), the position being dictated on the vertical axis by the lightness of the product. Most whiskies can be expected to fall within the red–yellow quadrant of colour space.

(Figure 5.9b). The closer products are in this three-dimensional space, the closer they are considered to be in terms of colour perception.

The presence of taints should be avoided wherever possible. Tainting species are generally characterised by their intense flavour activity, although they are usually not considered harmful, beyond damaging the flavour of the product. Several taints can be found in whiskies (Table 5.1). One of the more common taints is trichloroanisole (TCA). Perceptible as a musty taint at levels of less than 1 μg L^{-1}, most occurrences are commonly thought to derive from microbial methylation of trichlorophenol, which in turn was a contaminant in the wood preservative, pentachlorophenol (PCP). Even though the use of PCP for wood preservation has been discontinued, residues of tricholoanisole find their way into products when there is some, even indirect, contact of materials or plants

Table 5.1 Selected undesirable adventitious compounds that can occur in whisky.

Compound	Structure	Negative impact	Amelioration
Ethyl carbamate (EC)		Suspected carcinogen	Fresh copper; select low precursor malts
N-nitrosodimethylamine (NDMA)		Suspected carcinogen	Low NO$_x$ and/or SO$_2$ in kiln gases
Naphthalene		Mothball off-flavour	Care with storage of final product
Geosmin		Earthy off-flavour	Moisture management during grain storage
Trichloroanisole (TCA)		Musty off-flavour	Quality management of wood and cork

with contaminated wood. TCA is also the primary flavour impact character of 'corked' taint, as it can arise from corks used in the wine and spirit industries.

Geosmin (geo = earth, osme = smell) contributes a contaminating earthy palate to whiskies and appears to originate from fungal activity on grain that has been stored in damp conditions, allowing the offending fungi to grow. Interestingly, geosmin seems to be able to survive fermentation, distillation and maturation to find its way into the final product. Although it is also commonly found in water, it is not always possible to rule out the possibility of water contamination. Occasionally, whisky stored under poor conditions can develop a naphthalene (mothball) taint. It is not clear where this originates from (although it is produced by certain fungi), but nonetheless some customer complaints can be traced back to the presence of μg L^{-1} levels of naphthalene.

As with all foods and drinks, it is necessary to pay attention to potential safety and integrity issues. The two issues of particular importance are N-nitrosodimethylamine (NDMA) and ethyl carbamate (EC; Table 5.1). NDMA is both toxic and a suspected human carcinogen and is derived from nitrogen oxides generated from fuel used in the kiln drying of malt. The levels of NDMA are now well under control, due the presence of sulphur oxides in kiln heating gases and the development of a simple method, using a thermal energy analyser, to monitor products and raw materials.

EC caused substantial interest and concern when it was discovered. Whilst not acutely toxic, it is considered to be a probable carcinogen and therefore comes under legislative control in many countries. It was found that EC forms post-distillation due to the reaction between trace cyanide and cyanate precursors, in turn derived from the naturally-occurring cyanogenic glycoside, epiheterodendrin, which is found in the acrospires of malted barley. A substantial research effort found that the levels of the glycoside were variety dependent, so this has become a specification by which new malting varieties are selected. Additionally, the presence of an additional copper surface effectively sequesters measurable cyanide, so that there are now approaches to effectively manage EC in the final spirit. EC can be measured by gas chromatography with a nitrogen-specific detector; although, GC–MS, with selected ion monitoring at *m/z* 62, is now increasingly common. It is worth pointing out that the widespread occurrence of EC in foods and drinks means that the legislative focus is on reducing the body burden of EC, rather than targeting particular products or groups of products.

5.6 SENSORY EVALUATION OF NEW MAKE SPIRITS AND WHISKY

From what has been discussed before, it is clear that sensory evaluation plays a business-critical role in the production of new make spirit, during blending and for final product release. Similarly, some form of sensory evaluation is required to ensure, for instance, the integrity of casks (especially in terms of potential taints) and the sensory quality of raw materials. Because of the high alcoholic strength of new make spirits and final products, sensory evaluation is usually restricted to sniffing, or 'nosing'. If the sample is to be tasted, then it is usually circulated around the mouth and may be swallowed or not. For some sensory attributes that are due to the presence of compounds of lower volatility, such as the expression of wood- and peat-derived flavours, tasting can be a more sensitive arbiter of such flavour qualities.

Generally the high alcohol strength of spirit and final product precludes detailed sensory evaluation of the samples as is, and they are generally diluted with chlorine-free tap water or bottled mineral water prior to evaluation (Table 5.2). This also 'opens up' the spirit by increasing the flavour activity of the contained flavour-active congeners. The typical glass is around 150 mL, tulip-shaped and

Table 5.2 Typical spirit and whisky dilution regimes (adapted from the Institute of Brewing's Sensory Analysis Manual).[3]

Sample	Sample strength (%ABV)	Whisky + water (v/v)	Testing strength (%ABV)
Bottled whisky	40	1 + 1	20
Matured ex-cask	55–63	1 + 2	18–21
New make post-reduction	63–68	1 + 3	16–17
New make malt ex-distillation vat	68	1 + 3	17
New make grain ex-distillation vat	94	1 + 4	19

may or may not be fitted with an inverted domed lid. The samples of around 40–50 mL are usually presented in clear glasses; however, where colour might influence the assessors, dark glasses or testing under red lights are more appropriate conditions.

The sensory tests applied are dependent on the question asked. For instance, to establish the similarity or otherwise of two products, a triangle or duo–trio test is appropriate, where individuals are asked to match two samples and identify the 'odd one out'. For the triangle test, there is a one in three chance of choosing the correct answer by chance so, typically, 24 or more assessments are required to determine whether there is a statistically-significant difference or not. This requires around one bottle of 40% ABV whisky, so occasionally the much more challenging two-out-of-five test is used. Here, fewer assessors are required as there is only a one in 20 chance of guessing the right pair of samples.

If there are differences, or a more detailed evaluation is demanded, then some form of qualitative or quantitative descriptive analysis can be applied. Assessors typically are asked to judge the aroma (and perhaps taste and aftertaste) of products and note down their views and/or scores either on paper or a computerised system. Such evaluations are not straight-forward to perform and it takes a significant amount of training to become competent. The terminology typically used for a checklist (Figure 5.10) tends to be more high level in terms of requiring integration of several flavour attributes to make a judgment on each. Flavour profiles on the other hand (Figure 5.11) tend to be based more on chemical terminology derived from variants of the whisky flavour wheel (Figure 5.12). The flavour wheel attempts to place similar sensory terms in close proximity to each other. This similarity might be in terms of human association, such as cut and dried grass, or more mechanistic, such as associating all the taste attributes (*i.e.*, sweet, salt, sour, bitter) together. Clearly there is an element of subjectivity in the choice of descriptors and their positioning on the wheel but, nonetheless, it is a useful guide for trainees and experienced sensory assessors alike.

The subsequent analysis of such data is a question of considerable debate. Sensory data is rarely normally distributed and continuous; however, nonetheless, parametric tests, including multivariate tests such as principal component analysis (PCA), are routinely used.

WHISKY FLAVOUR CHECK SHEET

Name... Date... Time..
.   .   .

Sample
details...

Appearance.................................. Colour... Clarity..
..   .   ..

Aroma	Comment/level	Satisfactory (yes/no)
Total intensity of aroma		
First impression of aroma		
Whisky identity*		
Nose effect/feel (pungency *etc.*)		
Fruity/estery/fragrance		
Cereal like/feints character		
Sweet associated character (non-ester)		
Peaty/smoky character		
Other notes (specify)		
Off-notes (specify)		
Flavour by mouth	**Comment/level**	**Satisfactory (yes/no)**
Total intensity of flavour		
First impression of flavour		
Whisky identity*		
Mouth effect/feel (warming *etc.*)		
Primary tastes (sweet, sour, bitter)		
Fruity/estery/fragrance		
Cereal like/feints character		
Sweet-associated character (non-ester)		
Peaty/smoky character		
Other notes (specify)		
Off-notes (specify)		
After-taste (1 minute after sampling)	**Comment/level**	**Satisfactory (yes/no)**
Character		
Linger		

* Holistic appraisal of the whisky, including trueness-to-type

Figure 5.10 A typical sensory check list for assessing whiskies (based on the Institute of Brewing's Sensory Analysis Manual).[3]

A visual, although not necessarily statistically rigorous, way of presenting flavour profile data is to generate a radar or star plot (Figure 5.13). This can be particularly helpful when comparing the sensory profile performance of a small set of products.

Whisky/spirit aroma assessment

Name: **Product code:**

Instructions

Nose the spirit as instructed, and score the attributes below on a scale of 0 - 10. The inclusion of comments is strictly optional.

Aroma descriptor	Score	Comments
Pungent		
Smoky/peaty		
Medicinal		
Malty		
Grainy		
Grassy		
Floral		
Fruity		
Solventy		
Oily		
Smooth		
Sweet		
Buttery		
Nutty		
Woody		
Vanilla		
Spicy		
Mould/musty		
Sulphury		
Rancid/cheesy		
Soapy		
Catty		
Sour/acetic		
Preference rating		

Figure 5.11   A typical flavour profile for assessing whiskies. Assessors are asked to nose the spirit and rate the intensity of the indicated sensory attributes on an unstructured 0–10 scale.

Figure 5.12 A simplified whisky flavour wheel, showing primary (bold) and secondary descriptors. Further refinement, using a tertiary set of terms, can be used to further resolve the secondary terms. Terms that might be considered to be similar tend to be co-located. (From Ref. 4).

5.7 COUNTERFEIT DETECTION AND BRAND PROTECTION

The value of many whisky brands is substantial and has attracted the attention of criminal elements wishing to take advantage of the market position that these products enjoy by deliberate substitution of the original product with a cheaper variant. This can be considered to be at two levels:

1. Brand authenticity: is the liquid/package what it purports to be?

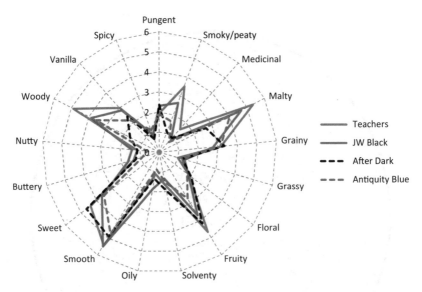

Figure 5.13 A radar plot of a selection of whiskies derived from the flavour profile shown in Figure 5.11. Common features include low scores in the 'flavour-negative' part of the plot (*e.g.*, mould/musty, catty, cheesy).

2. Generic authenticity: can the liquid be considered to be whisky?

Brand authenticity is the more challenging issue to address. Typical scenarios include the partial or total substitution of the original brand with a cheaper alternative, whilst too high or too low ethanol concentrations suggest some form of impropriety – the former suggesting an admixture with, say, neutral spirit, whilst the latter implies over-dilution. Such activity defrauds the consumer, the brand owner and, depending on the situation, the bar owner. How suspected cases are dealt with depends on where the problem arises. In the UK, for instance, the most recent legislation (the 2009 Scotch Whisky Regulations) appoints the HMRC as the competent authority for the verification of Scotch whisky. Suspect samples are analysed by the government or brand owner laboratories to try and establish whether or not there has been any criminal activity. Analytical confirmation of authenticity can be problematic, although an analysis of the major volatile congeners and alcohol strength is often informative. The presence of atypical levels or unusual ratios of non-volatile congeners can indicate that maturation has not been carried out conventionally and the presence of artificial flavours can, at least for some whiskies, reflect some form of adulteration.

Generic authenticity issues can also be addressed by considering the levels of ethanol and the major volatile congeners. If the product is not whisky then such analyses should quickly reveal problems and high levels of congeners, such as methanol, can indicate that some of the ethanol comes from an unacceptable – perhaps non-grain – source.

5.8 PRODUCT STABILITY

Once bottled, whisky is considered to be rather stable and indeed, when well-sealed, the concentration of ethanol does not change significantly over time. More commonly, whiskies can develop precipitates, particularly if they are not chill-filtered or if they are stored in cold conditions for an extended period of time. Unlike a change in flavour nuance, the presence of precipitated solids is unambiguous and a more easily shared experience and so is often an area of focus for whisky producers. In terms of changes in the liquid, water can migrate preferentially through PET bottles and result in an elevated ethanol content. It is possible to detect small increases in acetaldehyde and ethyl acetate in older whiskies in intact bottles. It is not clear what these increases are due to but it is highly likely that oxygen in the headspace of the bottle at bottling and any that can diffuse across the closure will cause some oxidation of ethanol. If this oxidation proceeds beyond acetaldehyde to acetic acid then this will disturb the ester/alcohol equilibrium resulting in ethyl acetate formation. New make spirit can contain 1–5 mg L^{-1} copper and perhaps residual copper in the bottle feasibly could catalyse the formation of singlet oxygen to accelerate ethanol oxidation. The exposure of whiskies to light, especially if they are colour-adjusted with caramel, is likely to cause the colour to fade. Thus, storage in the dark is advisable to retain the colour profile of the product.

5.9 IN CONCLUSION...

The definitions and permitted compositions of whiskies around the world are dependent on the legislation that pertains to the specific

sales markets and allow for a wide range of permitted formulations
that can be called whisky, a situation that leads to vigorous debate
concerning the definition of whisky. Most of the final products that
are called whisky have essential features. They are often grain-
based, although some whiskies can be legally produced without the
requirement for grain. Generally, whiskies are subject to a maturation process in the presence of oak wood. All matured whiskies
fall into the category of 'brown spirits'.

The value of the whisky market worldwide has encouraged the
growth of an unscrupulous counterfeit industry, which potentially
compromises consumer protection. Nevertheless, these counterfeiting operations are being robustly challenged by the actions of
organisations, such as the Scotch Whisky Association.

NOTES AND REFERENCES

1. Of course, this is not to say that single malts cannot share the
same qualities!
2. A. S. Gray, *The Scotch Whisky Industry Review*, 35th edn,
Sutherlands, 2012.
3. Institute of Brewing, *Sensory Analysis Manual*, Institute of
Brewing, UK, 1995.
4. K.-Y. M. Lee, A. Paterson and J. R. Piggott, *Journal of the
Institute of Brewing*, 2001, **107**, 287–313.

CHAPTER 6

Marketing and Brand Development

The 'classical' academic approach to marketing defines the activity as *"the set of human activities directed at facilitating and consummating exchanges"*.[1] It is not entirely fatuous, however, to suggest that such an all-encompassing definition might include much other human activity involving exchanges, some of which at least take place in dimly-lit spaces, enhanced by the consumption of whisky!

Be that as it may, at the more practical level of the marketing practitioner, management theory has talked of the 'four Ps' of marketing since 1960: product, price, promotion and place. As will hopefully become clear, each of these Ps embraces a series of more or less loosely related disciplines with a multitude of specialist tasks and sub-tasks. Such is the complexity of the marketing challenge that practitioners will experience considerable variation between the approach adopted by smaller companies, where the marketing manager will of necessity be something of a generalist, compared to larger concerns, where specialism and the employment of highly-focused external agencies and consultancies is more prevalent. It is also the case that the budget allocated to any given brand or enterprise will, to a significant degree, determine the level and intensity of marketing input.

More recently, Lauterborn[2] and others have proposed replacing the four Ps model with four Cs: consumer (for 'product'), cost

The Science and Commerce of Whisky
By Ian Buxton and Paul S. Hughes
© Buxton & Hughes, 2014
Published by the Royal Society of Chemistry, www.rsc.org

('price'), communication ('promotion') and convenience ('place'). However, whether described as Ps or Cs, these dimensions are then broadly encompassed in an over-reaching 'brand positioning statement', sometimes graphically represented as the 'brand architecture'. Concepts, such as 'brand DNA', have also been developed in an attempt to capture the history of a brand's development and thus facilitate the incorporation of this marketing heritage into current activity. The brand DNA concept assumes particular significance in a market, such as whisky, where considerations of provenance and heritage are important factors for many consumers.

It should be noted that this commentary is, inevitably, somewhat broad-brush as it attempts to cover in one brief chapter what has become a branch of management science, with complete undergraduate courses of three and four years' duration devoted to its study, not to mention post-graduate and MBA studies. It should also be made clear from the outset that this chapter does not attempt to address the sales and distribution functions, or the process of budget setting and resource allocation in marketing planning, except for very peripherally. Again, these activities are large and significant, more than justifying study on their own account.

6.1 PRODUCT

It might be simplistically imagined that the product is a given, determined by the production function and provided to their marketing (and sales) colleagues as a fixed and unvarying constant that is not to be questioned. While this may once have been the case within the distilling industry it is certainly no longer so and much time, attention and energy is devoted to creating new products that it is considered the market will adopt with enthusiasm and to fine-tune existing products to maintain their competitive edge, even if the brand mythology is one of unchanging consistency.

It is in the nature of whisky that some variability in the product is inevitable. While, for many years, the skill of the blender has been to minimise these variations, even long-established brands do evolve over time in response to changing consumer tastes and varying regional preferences, let alone the availability of specific whiskies in the blend.

There is some evidence to suggest that blended Scotch whisky was peatier in flavour in the early part of the 20th century; it then

evolved to offer a smoother, lighter taste (*e.g.*, with the introduction of brands, such as Cutty Sark and J & B during the 1920s,
principally for the American market), while in recent years more
fully flavoured and smoky variants have been introduced
(*e.g.*, Johnnie Walker Double Black and The Black Grouse).

This is most clearly seen in the single malt sector, where strongly
flavoured peated malts were all but unsaleable as late as the 1980s
(the reason, in part, why a number of Islay distilleries were closed
or mothballed) but have subsequently made a dramatic recovery.
Today, a plethora of different ages and finishes are offered to meet
the requirements of the enthusiast consumer. Indeed, as argued in
chapter 5, some smaller craft distillers have made a virtue of the
constantly changing nature of their product and, in doing so,
turned the orthodoxy of mass-market brand marketing, where
consistency is paramount, on its head.

There are, of course, limits to the extent to which a well-known
brand can alter its product without alienating loyal consumers. The
Johnnie Walker family, for example, has a distinctively smoky
note, which defines the house style, while a brand such as J & B is
defined by a lighter, grainier style. Thus, any new Johnnie Walker
expression will have, at its heart, some smoky whiskies and the
regular Walker consumer would be surprised and disappointed if
this distinctive flavour note was absent. The house style of the
whisky may thus be considered part of the brand DNA.

However, at least for major brands, even given these parameters,
a considerable amount of research and development will go into
the creation of a new product. Initial work in the laboratory or
sample room will provide marketing with a number of options that
can be exposed to both financial analysis and trade and consumer
research. Assuming that the proposed new product can be delivered at the required price point and target profitability, then
product variants can be presented to groups of consumers carefully
selected to be representative of the target market and their reactions fed back to the laboratory in an iterative process (such
research is known as qualitative research and the usual technique
employed here is the focus group).

At the same time, the consumer group may be asked to look at
and critique packaging designs or advertising executions and proposed pricing; they may also discuss how the whisky will be purchased and consumed and consider how the introduction of this

new variant affects their overall impression of and relationship with the brand.

Larger brands with more substantial budgets may well maintain long-term quantitative tracking studies which, as the name suggests, aim to track changes in the consumer's image of and attitude to the brand.

In recent years, the whisky industry has seen a considerable number of new products brought to market. Even in the relatively constrained Scotch whisky sector there has been innovation in cask finishing and the development of premium and super-premium brand expressions. The market's acceptance of extra-aged whisky, *i.e.*, 25 years and older, has been a phenomenon that would have surprised distillers and marketers as recently as 20 years ago, when the prevailing orthodoxy was that spirit of this age was 'over-aged' and likely to be excessively woody in flavour. The premium pricing associated with whisky in this category has been welcomed by the industry (and by retailers) both for the handsome profit margins it delivers and the associated trickle-down of positive imagery for the brand.

Even more recently, however, as the availability of aged whisky has declined, new product development has reflected this with the launch of a number of premium whiskies that do not carry an age statement, so-called non-age statement (NAS) products. Here, the brand owner argues that other factors, such as cask selection and blending, are more important than age alone.

The reader will appreciate also that even in a product as apparently straightforward as a new aged variant of a single malt there is scope for product development (assuming that the distillery has a range of cask types of the required age). By altering the mix of casks in the final vatting, through the selection of the bottling strength, by deciding to chill-filter and by adding spirit caramel or not the product character can be designed to meet a price point and appeal to a particular consumer group – thus, only with a single cask bottling can it be truly said that the product is fixed and given, wholly determined by the production process and free from manipulation by the marketing function.

6.2 PRICE

The final price of a bottle of whisky is a function of a number of variables, not all of which are under the control of the brand owner.

A price, ex-distillery, is determined by the brand owner, who may also control the margin taken by the relevant distributor and wholesaler (even if the brand owner does not actually own the distribution company the margin will often be contractually fixed in the supply agreement or by informal discussion in light of market conditions). However, retailers in both the on- and off-trade are generally free to fix their own margins within the constraints of the competitive pressures in the market and, while the brand owner or distributor may attempt to influence this by negotiation, promotional offers or portfolio management, they cannot control retail pricing. Indeed, in some countries, most notably the USA, distribution arrangements for alcohol are consciously designed to distance the brand owner from the retail operation.

In markets where a government monopoly on retail distribution is maintained, a policy of so-called 'social reference pricing' may be applied. Essentially, this consists of keeping retail prices artificially high in the belief that this will discourage excess consumption – considered to be socially desirable. At the time of writing, there has also been much debate and some political controversy over proposals for a minimum retail price per unit of alcohol in the UK, set through the mechanism of taxation. Proponents of such an approach point to the societal impacts of excessive alcohol consumption to justify the cost to all drinkers, moderate or otherwise, and it is explicitly driven by a desire to effect behavioural change in a particular group in the population.

A factor of considerable importance, of course, is taxation – both duties on alcohol *per se* and sales taxes. Additionally, in a number of markets, whisky faces tariff barriers on its original importation. India, for example, levies import duty at 150% before the 28 individual states add their own local duties, which some commentators have suggested are not entirely uninfluenced by the presence in any given state of a part of the sizeable Indian distilling industry. Such costs are, of course, entirely beyond the control of individual brand owners, who may be only slightly consoled by the thought that all their competitors face the same hurdle. Industry bodies, such as the SWA and the Distilled Spirits Council of the United States (DISCUS), are constantly lobbying at government level to lower tariff barriers and have enjoyed some success in recent years.

As an aside, India is second only to Egypt in tariffs. Egypt's tariff is 300% if the product is going to 'tourist establishments' (*i.e.*, hotels); otherwise, duty is levied at a remarkable 3000%. Not surprisingly, Egypt does not feature in the leading export markets for whisky.

One result of uneven levels of taxation has been the development of the 'grey market' or parallel imports, involving the legitimate purchase of products in one country at a price below that of a second country and then importing the products into the second country, where they are sold below the local market price. Such a practice, though frowned on by brand owners seeking to protect their local distributor (or maintain profit margins), at least involves genuine products.

Not so is the practice of counterfeiting, where an inferior product is passed off as the genuine article (Figure 6.1). This may involve a cheaper whisky being sold as something of higher quality,

Figure 6.1   Examples of imaginatively named and packaged counterfeit 'Scotch whisky' detected by the Scotch Whisky Association. Courtesy SWA.

as in the notorious Pattison case of the late 1890s, in which Irish grain whiskey was vatted with a small amount of single malt and sold as 'pure Glenlivet', or even more perniciously, where a locally-produced spirit of dubious origin is bottled and sold as a well-known brand. The larger brand owners and industry associations continually police this type of activity but, notwithstanding their vigorous efforts, the easy profits to be made continue to attract miscreants and malefactors in search of the proverbial 'quick buck'.

While counterfeit products may be susceptible to chemical analysis,[3] this requires both access to laboratory facilities and time for the analysis to be performed. It is not, therefore, a realistic tool in the field. However, more recent work at University of Strathclyde University,[4] University of St Andrews[5] and the University of Leicester appears to hold out the promise of a variety of approaches to this problem, which has been estimated to cost the Scotch whisky industry over £500m in lost sales in the Far East alone. While it could be argued that the 'lost sales' argument is somewhat specious and any value figure artificially inflated, in that purchasers of very cheap whisky are unlikely to be consumers of the genuine article, the issue of reputational damage to the brand or the whisky category as a whole arising from the presence of counterfeit products is considerably more serious.

Lest the problem of counterfeiting be thought to apply only to developing markets and it should be noted that high levels of taxation in the UK have contributed to a counterfeiting problem there. As recently as June 2012, trading standards officers from Rochdale Borough Council seized 13 bottles of High Commissioner Whisky and 18 bottles of Glen's Vodka, all suspected to be counterfeit, from an off licence in the borough and it was reported that, in recent months, officers had seized a total of 192 bottles of suspected counterfeit alcohol –including 126 bottles of whisky.[6] (See also Box 6.1 for the case study on Lowland Glen and Glencarnie.)

By way of illustration, Figure 6.2 illustrates the contribution of alcohol duty and sales tax (VAT) to the retail price of a typical bottle of standard blended whisky in the UK in October 2012. It will be immediately seen that of the price paid by the consumer, the major beneficiary is the government – some 80% of the £11.27 shelf price! However, such is the structure of the taxation system that the percentage accounted for by taxation actually declines as the retail price increases.

BOX 6.1   CASE STUDIES: COUNTERFEITING

Industry bodies, such as the Scotch Whisky Association (SWA) maintain a considerable effort in identifying, tracing and, where appropriate, pursuing legal action against counterfeit products.

At times, such as their extended but ultimately unsuccessful legal action against the Canadian distillers of Glen Breton, this may seem unduly pedantic but there are vitally important commercial and legal considerations to be taken into account.

In total, the SWA Council has authorised legal action against over 1000 brands and nearly 3000 trademarks worldwide have been opposed. Today, at any one time, up to 70 different legal actions around the world are being pursued. Such legal work is often a long term effort, with the longest case thought to be a trademark opposition in India that began in 1964: 'Highland Chief' was still under appeal 27 years later when agreement was reached.

Less seriously, the association's lawyers have found themselves pursuing a Filipino producer who had claimed to invent unique technology that meant 35 minutes of vigorous shaking in Manila equalled 35 years of maturation in Scotland. There was also an Indian case, where a producer argued he had named his brand 'Scotch Terrier' out of love for a pet rather than any attempt to misuse Scotch whisky's international reputation.

Some recent examples of the SWA's work illustrate the scale of this continuing problem. The SWA receives reports from member companies of suspicious activity which are then energetically followed up by an in-house team of 5 lawyers.

RED BIRD

RED BIRD was seized by HMRC at Dover and analysis indicated it was neither Scotch whisky, nor whisky in terms of the definition of whisky in European law. The product was traced to Spain, where the SWA undertook successful criminal proceedings.

GRANT'S REGAL

A German businessman contacted the SWA to express his doubts that a canned product called GRANT'S REGAL

Blended Scotch Whisky, being sold in the Middle East, was genuine. Analysis concluded that the contents were not Scotch whisky as the spirits had not been produced from cereals and had not been matured. Enquiries made by the Scotch Whisky Association showed that the product was being transported to Turkey from Europe for distribution to the wider Middle East region. Applications were filed with customs in Belgium, Austria and Germany for seizure of any shipments. The association's application to Austrian customs was successful and a shipment was seized, leading to legal proceedings against the Austrian company concerned. Company records revealed more than 15 million cans of fake Scotch whisky had been exported over a 3 year period.

LOWLAND GLEN and GLENCARNIE

These two products were found on sale at car boot sales and from licensed premises in the UK. Although not described as Scotch whisky, they were labelled with names associated with Scotland and other misleading indications of origin. The SWA traced the source of the product to a French company and a raid was carried out, revealing that some 2.5 million bottles of these brands had been smuggled into the UK over a 2 year period.

The contribution of the Scotch Whisky Association to this case study is gratefully acknowledged.

In general terms, brand owners pay considerable attention to the price of a basket of their competitors' products – the marker brands that they consider to be equivalent to their brand in positioning terms or which lead the sector in which their brand is sold. Achieving the 'right' pricing for any product is a critically important element of the marketing mix and vital to achieve optimum brand profitability.

Consumers, especially in developing markets, may attach high significance to price as a mark of quality and status. On occasion, pricing of one brand variant can be used to increase sales of another, as witnessed by the duty free pricing strategy of Johnnie

Figure 6.2   The contribution of alcohol duty and sales tax (VAT) to the retail price of a typical bottle of standard blended whisky in the UK in October 2012. Courtesy SWA.

Walker's Blue Label King George V Edition *versus* its stable mate, Johnnie Walker Blue. The KGV Edition is typically priced at around three times the price of Blue and, anecdotally, it is reported that prominent display of the higher-priced whisky results in an increase in sales of Blue of around one third compared to its display in isolation − the image of the KGV Edition reflecting its halo effect onto its junior expression.

Finally, much prestige is invested in achieving the ultimate in super-premium pricing, as witnessed by the kudos that rival distillers attach to having the most expensive bottle in the world (Figure 6.3). A number of bottles of Scotch whisky have been released in the recent past with prices in excess of £100,000, although the limits and size of this market remain to be fully tested.

Figure 6.3 Following its sale at Sotheby's, New York, for $460,000, the Macallan Cire Perdue has been recognised by the Guinness World Records as the world's most expensive whisky sold at auction.

Price is a key determinant in luxury marketing, especially where supplies of the product are deliberately restricted by means of limited editions and similar techniques. Provided that the brand's reputation can sustain it, a deliberately high price where the brand is managed for value (profit) rather than volume may actually act to reassure rather than deter consumers. The very existence of a high price can thus act to make the brand more aspirational and therefore increase desire for it. Even in conditions of general economic recession, the above appears to hold true and the industry's recent experience has been of whisky pricing particularly at the luxury end of the market apparently immune to wider economic forces.

Some commentators have thus suggested that certain brands of whiskies could be purchased and stored as part of an alternative investment strategy. At the time of writing, the long-term performance of such a portfolio remains largely a matter of conjecture, however; although at least one specialist 'investment' vehicle has been launched claiming to achieve returns of 15% capital growth annually.

6.3 PROMOTION

Promotion (or 'communication') reflects a diverse range of activity ranging from public relations, sponsorship, direct mail and advertising through to the more recently developed channels of social media. All represent 'touch points' for the brand, where a particular target group, be that trade or consumer, receives (and, in the case of social media, may transmit) messages about the brand.

The aim of the marketing professional is to ensure that all such communication is consistent with the brand's overall positioning and presents a coherent image of the brand. While that may seem simple, straightforward and logical in theory, in practice, few whisky brands manage to achieve this on a global basis as, all too often, different messages continue to be communicated in different geographical markets.

This is very much a function of the historical development of the whisky market. The early sales pioneers were concerned primarily in shipping their product to international markets and the job of communicating the brand image was largely left to the relevant distributor, who might receive a sales visit from the principal once a year or less. Historically, such distributors were independent firms, not controlled by the brand owner, and they were permitted (or abrogated to themselves) considerable latitude in developing communications that they deemed appropriate or relevant to their local market conditions. Accordingly, quite different and sometimes conflicting images could, on occasion, be developed for the same brand in different markets.

Increasingly, however, as the marketing function has grown in strength and consolidation of the industry has concentrated ownership, international distribution has come under the direct control of the brand owner and there has been a concerted drive to establish internationally consistent positioning, at least for major brands. Perhaps the best example of this is Diageo's Johnnie Walker brand (see Box 6.2 for further details). Two further relevant brand case studies for a major blend and a single malt are illustrated at Boxes 6.4 and 6.5 respectively.

This process has been accelerated by the growth in international travel by consumers (hence the importance of the tax free channel to sales), by the rapid increase in the use of the web by consumers and by the increased presence of 'brand ambassadors' in all major global markets.

BOX 6.2 CASE STUDY: JOHNNIE WALKER BLENDED SCOTCH WHISKY

Arguably, no other company can claim such a distinguished and sustained history as blenders and distillers of Scotch whisky as Johnnie Walker, today part of Diageo. From its original modest origins (in 1820) as an 'Italian grocers' in Kilmarnock in the west of Scotland, the company remained in family ownership until 1925, when it became part of the DCL.

Johnnie Walker Red Label was the first truly global brand of Scotch whisky, selling in over 120 world markets by the end of World War I. Today, it is the largest selling blended Scotch whisky in the world, the flagship variant for a brand that has revitalised itself over the past fifteen years with its global 'Keep Walking' campaign, Formula One sponsorship and exceptional product innovation and premiumisation.

Unlike some other whisky families of their day, the Walkers were not international playboys but, in a very Scottish way, maintained an understated, almost austere demeanour re-investing the majority of their profits back into the business. They were, first and foremost, focussed on business success.

It was a success story built on some simple principles. "*So far as quality is concerned, nothing in the market shall come before it*", wrote Alexander Walker, the second generation of the family. Quality credentials were underpinned by a string of medals from international exhibitions, critical then (and today) in establishing the pedigree of brands as they went into new markets. And Walker's mantra, "*our blend shall not be beat*", was also reflected in the careful way the business built up stocks, starting with a few casks of mostly west coast whiskies in the original grocer's shop, (giving the Walker brands a distinctive smoky character, which they retain to this day).

By the 1880s, they occupied substantial premises in Kilmarnock and, seeing the necessity to, build a vertically in-tegrated business of purchased distilleries, such as Cardhu. Today, the Walker blending team have access to over seven million casks of maturing Scotch whisky from Diageo's sub-stantial operations and the industry practice of purchasing new fillings and exchanging mature stock.

The company was an early pioneer of marketing: trade-marking the label for Old Highland Whisky in 1867, a distinctive black and gold design carefully placed on the bottle at an angle of 24 degrees. With the subsequent adoption of the iconic square bottle in the 1870s it has since developed into one of Scotch whisky's best-known and most recognisable designs. Such design innovations were key to the brand's success and, in an era before containerised shipping, the square bottle proved both economic to package and ship and less vulnerable to breakage, thus increasing the firm's relative profitability.

A young and dynamic third generation of the family consolidated their father's work but also embraced change reinventing Old Highland as 'Black Label' and introducing alongside it 'Red Label'. In the 1880s, they inherited a mostly UK-focussed business, with limited export trade. By 1920, Walker could lay claim to being the world's first global drinks brand, exporting to over 120 markets, assiduously developed by careful selection and appointment of strongly incentivised distributors.

Today, of course, Diageo directly controls almost all of its international distribution, except where prevented by local legislation, as a further step towards a completely integrated business and seamless supply chain.

Walker's had been slow to grasp the nettle of mass advertising but eventually embraced it both fully and well. A series of leading agencies and graphic designers have, for over 100 years, turned artist Tom Browne's original cartoon of a 'striding man' figure into one of the world's most recognised brand icons. Today, this anchors the brands' 'Keep Walking' campaign, inspiring personal progress in its consumers around the world.

Walker's global campaigns are noted for a high degree of central control, ensuring great consistency of both image and message. Walker's are active in virtually all media, including above- and below-the-line activity, sponsorship and PR. They have actively embraced social media and web-based marketing, employ a considerable team of brand ambassadors and have created notable brand experience destinations, such as the Johnnie Walker House in Shanghai and Beijing.

In recent years, there have been a number of premium and super-premium line extensions, generating significant gross margins and much favourable PR, such as Johnnie Walker Double Black, Blue Label King George V Edition and the extremely limited Diamond Jubilee Edition (60 bottles only to commemorate the Diamond Jubilee of Queen Elizabeth II at £100,000 a bottle). Such line extensions are often employed tactically in selected markets or channels, but all rigidly adhere to the overall brand guidelines and positioning.

Today, the Walker range of blended Scotch sells in excess of 16 million cases and is the dominant global brand leader.

Key success factors:

- Long-term consistency of ownership, with an emphasis on sustained quality and a distinctive and consistent blending style
- Access to unrivalled stocks and high quality international distribution – group ownership ensures the interests of the distiller, marketer and local distributor are fully aligned and all profit retained within the parent
- Consistent re-investment in innovative packaging and advertising
- A compelling brand heritage and provenance
- A highly successful, well targeted and profitable strategic programme of line extension built on strong brand values and deep understanding of the consumer's needs and values

The contribution of Diageo PLC to this case study is gratefully acknowledged.

6.4 PLACE

In the 4Ps model, 'place' was synonymous with distribution or the place(s) where the consumer would be expected to find (and buy) the product.

Conventionally, in the beverage alcohol market, this is separated into on- and off-trade channels, referring to whether or not the product is consumed *on* (pub, hotel, restaurant, *etc.*) or *off* the premises (liquor store, supermarket, duty free outlet, *etc.*). Such premises are generally highly regulated with strict controls on opening hours, limitations on the age of the purchaser and, in some cases, the other goods that may or may not be retailed there.

Such restrictions have become somewhat blurred with the advent of on-line shopping and specialist on-line retailers have assumed considerable significance in the sales of speciality whiskies, such as collectable single malts. 'Convenience', Lauterborn's suggested replacement term for 'place', attempts to acknowledge the profound impact of the internet on the consumers' knowledge and purchasing behaviour.

Note, however, that in some markets, such as certain US States, Canada, the Nordic countries and elsewhere, liquor retailing is controlled by the government and operated as a closely-controlled monopoly.

Increasingly, also, 'place' has come to refer to the distillery itself or other 'brand place', such as pop-up brand experiences, cocktail bars and sampling events, such as Whisky Live.

6.5   BRAND POSITIONING

The summary of the 4Ps is expressed in the brand positioning statement (sometimes referred to as the 'brand architecture'), which attempts to capture the values, features and benefits (emotional and functional) of the brand. This is then further summarised into a 'brand truth' or 'brand promise' which, in theory, is used to guide all proposed marketing activity for the brand, which should be evaluated against this 'truth' prior to implementation. However, as noted above, in practice, many brands exhibit highly variable regional behaviour and cultural factors further complicate this attempt to simplify the brand's communications.

An example of brand architecture is shown in Figure 6.4 for Glenglassaugh, a small Highland distillery re-opened in 2008, and which relates to the positioning of the remaining stock of older whiskies acquired with the distillery and marketed by the new owners.

Perhaps the single key point to grasp in considering the marketing of whisky is that whisky, all whisky, is an entirely discretionary purchase for the ultimate consumer. No one *needs* whisky.

Furthermore, in a world in which there is – broadly speaking – no longer any bad whisky on sale, the decision to purchase is driven to a large degree by emotion and the consumer's relationship with the brand. Products can be copied but brands possess unique properties that cannot easily be replicated and, at their best, enter into the consumer's life, becoming part of the way in which he or she defines themselves to the world.

Glenglassaugh brand model

Brand promise	**A rare and enduring Highland malt: crafted by the elements and rediscovered for an elite few**				
Brand personality	Confident, even arrogant	Elegant	Classically stylish	Scottish but cosmopolitan	Individual and distinctive
Brand attributes	**A Hidden Gem**		**Exclusive and rare**		**Coastal Character**
	Tucked away on the Moray coast, Glenglassaugh has been re-discovered, and for the first time will be made available as a brand in its own right. Recognised only by the malt aficionado, the revival of this previously unknown distillery represents a unique opportunity for the discerning consumer.		Only tiny quantities of Glenglassaugh are available. Never actively "marketed" through traditional means. This is a Highland malt that will be sought out by those who are confident in their own judgement. Available initially in very limited quantities and made available on allocation to key markets and desired channels.		Established in 1875, Glenglassaugh has always been produced by generations of skilled distillers. The exposed, north-facing coastal site is honoured by time and favoured by nature for whisky production - the coastal elements ensure a distinctive whisky that is full of character and history.
Brand values	Exclusive luxury	Discerning	Individual and esoteric	Enduring	Something special

Figure 6.4   The Glenglassaugh brand model was developed to direct the marketing of the distillery's older (and more expensive) single malt whiskies.

Should the reader doubt the intensity of some fans' devotion to their chosen brand, then the illustrations in Figure 6.5 of real consumers who have freely chosen to 'brand' themselves at their own cost with permanent tattoos of their favourite whisky will dispel any doubts. At the extreme end of consumer behaviour, such fans can become quite compulsive in their devotion and the brand acquires cult status for them (see also websites, such as The Ardbeg Project, ardbegproject.com).

Such intensity of support, while at times overwhelming and even obsessive, is something that the average brand manager in consumer goods marketing would welcome and provides many niche brands, especially from smaller companies, with the opportunity to engage with highly committed brand supporters through social media at a low cost.

The phenomenon of collecting whisky, especially single malts, has similar roots and has grown significantly in recent years, supported by specialist auctions both physical and on-line. A number of commentators and brands have even suggested that whisky collecting can be pursued as an alternative investment opportunity, although this remains a controversial topic. There is no doubt, however, that astute brand owners have responded to the collecting

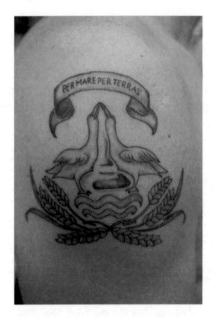

Figure 6.5   German whisky enthusiast, Norbert Leege proudly declares his brand allegiance to Glenglassaugh. Courtesy Glenglassaugh Distillery Co. Ltd.

craze with whiskies specifically designed for this market, where certain product and packaging cues, such as bottling strength, highly detailed labelling, a genuine (not printed) signature by the 'Master Distiller', numbered bottles and an accompanying leaflet or book, signal the 'collectability' of the release.

6.6   THE EXPERIENCE ECONOMY: BUILDING RELATIONSHIPS

The term 'experience economy' is derived from the title of an article and later book (1999) by the American management consultants, B. Joseph Pine and James H. Gilmore.[7]

Their theories have been influential, especially in tourism and in the development of so-called 'experiential marketing', seen in the whisky industry in the form of widespread product sampling at events, such as Whisky Live, the work of the many brand ambassadors now employed in the industry and, most dramatically, in the development of visitor centres and 'brand homes' at many distilleries (Box 6.3).

BOX 6.3   CASE STUDY: DEWAR'S WORLD OF WHISKY VISITOR CENTRE

The Dewar's World of Whisky visitor centre is based at the company's Aberfeldy distillery, in the Perthshire town of the same name. The distillery dates from 1896–1898 and was the first move into production by Dewar's, then an independent firm of blenders and bottlers. The company joined the DCL in 1925.

Until the Bacardi acquisition of John Dewar & Sons, Aberfeldy had been operated by the DCL and their successor, United Distillers & Vintners (UDV), essentially to provide fillings for the Dewar's White Label brand and other blends. Minimal attention was paid to the Aberfeldy single malt brand and, while visitors could be shown round the site, there were few dedicated visitor facilities. At most, a few thousand people visited the distillery prior to 1998 when Bacardi acquired the business. With it came an extensive brand archive covering packaging, correspondence and brand ephemera.

After a short review of their new sites, Bacardi was determined to make Aberfeldy distillery their principal whisky showcase. The distillery itself was upgraded and the site underwent an extensive renovation. The original malting had been closed when DCL centralised malting operations and, by 1998, was being used only for casual storage of surplus equipment. A feasibility study and business plan[†] identified the building as suitable for conversion to a visitor centre and specialist design consultants and architects were appointed.

Dewar's World of Whisky opened in April 2000 and was almost immediately acclaimed by *Whisky Magazine* as "*the ultimate Scotch whisky visitor centre*" and awarded 5* status by VisitScotland. The centre contains exhibits from the Dewar archive, now expanded by further purchases, used to provide visitors with "*a journey through time and spirit*" that employs the firm's history to understand the craft of blending.

In this, it was unique at the time as other centres concentrated exclusively on single malt whisky. Subsequently, the Famous

---

[†]One of the authors prepared the original feasibility study and led the design team for the project.

Grouse brand has opened a similar facility at the Glenturret distillery, although an emphasis on blending remains rare.

The tour, which is unusually self-guided, begins in a dramatic theatre-style room with an audiovisual presentation. Visitors then move through a series of room sets or tableaus, exploring different aspects of the company's history and brand personality. The tour concludes with an explanation of nosing and a tasting of the various whiskies in the café and shop area. Visitors can then continue to a warehouse display or an accompanied tour of the distillery. There are also spaces for brand training and induction. Exclusive 'distillery only' bottlings are available.

There is a heavy emphasis on interactivity throughout Dewar's World of Whisky. Around £1m was spent on the initial project and there has been subsequent investment. The most recently available figures show an annual attendance of ca. 35 000 with admission charges ranging from £7 (standard admission) to £75 for a private tour. Key success factors include:

- An innovative and unusual design concept with emphasis on blending and a high degree of interactivity
- High build quality and good maintenance has ensured a high standard of presentation is maintained. Enthusiastic staff members are encouraged to mix with guests
- An attractive site location in Highland Perthshire, close to major tourism flows
- Brand strength and appeal, especially to international visitors

It is perhaps not widely appreciated how recently the distillery visitor centre came to form such an essential part of the industry, such that it is now fairly exceptional for a distillery not to make some facilities available for consumer visitors, if only for restricted periods of the year.

The first formal distillery visitor centre was opened in 1969 by William Grant & Sons at their Glenfiddich distillery in Dufftown, Scotland as part of their pioneering promotion of single malt, reputedly at the then wildly extravagant cost of some £10,000. At the time, the industry tended to scoff at the idea of allowing the public into what was then a closely-guarded inner sanctum of the production team – today many distilleries are virtually an adjunct of the marketing function and it is not unusual to see expenditure in excess of £1m on visitor facilities.

BOX 6.4   CASE STUDY: CUTTY SARK BLENDED SCOTCH WHISKY

The original Cutty Sark blend was devised in March 1923 by Berry Bros & Rudd, an upmarket London wine merchant seeking to capitalise on the then current trend for cocktails by releasing a blend lighter in taste and flavour than the typical style of Scotch of the period.

By repute, the blend was created by Charles Julian but it is clearly documented that the iconic label was designed – probably over lunch on a napkin – by the well-known Scottish artist, James McBey. By 1938, sole responsibility for the supply of whisky for blending had passed to the Glasgow-based Robertson & Baxter (now part of The Edrington Group), although the brand was owned and managed by Berry Bros & Rudd.

Cutty Sark grew rapidly in the USA in the years following the repeal of prohibition, quantities of the brand having found their way there by various means *via* 'rum runners' out of Nassau during the 1920s and 1930s. The brand was associated with well-known bootlegger, Captain Bill McCoy, and his reputation for delivering only the 'real McCoy' to his grateful customers.

Largely due to the efforts of Berry's US distributor, The Buckingham Corporation, sales to the USA boomed and, by 1961, Cutty Sark was the number 1 blended Scotch whisky in what was then Scotch's largest export market. In fact, at that time, the USA accounted for around half of all Scotch's exports. Cutty Sark's sales then amounted to just over one million cases – the first Scotch whisky brand to achieve this level of sales.

The brand retained category leadership in the USA until the high watermark was reached in 1972 with sales of 2.75 million cases. However, Berry Bros were unable to sustain the considerable competitive pressure and marketing expenditure of J & B Rare, Dewar's White Label and, later, Johnnie Walker Red Label, and the brand started a long period of decline. Marketing in the USA appeared to lose focus and budgets were cut in the latter part of the 20th century, further contributing to the brand's decline.

By the early 2000s resources had been largely switched to developing Cutty Sark in Spain, Portugal and Greece and this initially met with some success, at least partly replacing lost US sales in smaller markets where Cutty could achieve and sustain some critical mass. However, this strategy proved highly vulnerable to the rapid and significant downturn in these markets following the financial crises of 2008 and onwards. Sustained depressed economic conditions in southern Europe proved highly damaging to Cutty Sark and, early in 2010, the brand was sold to The Edrington Group, who had been responsible for blending the whisky for more than 70 years.

The Edrington Group also own The Macallan and Highland Park distilleries and The Famous Grouse brand. As a private company, they are able to take a long-term approach to brand development and resist short-term pressures for profit maximisation. Since they have acquired Cutty Sark, the brand has been re-positioned as a gateway whisky designed to a younger, urban audience (albeit then supported with reference to its heritage); new packaging has been developed; the product line-up has been rationalised and award-winning new expressions introduced. Improved distribution arrangements have also been made. A £20m, five year marketing plan designed to increase sales by 40% has recently been announced. Key success factors include:

- Private ownership with long term focus
- Great brand heritage and 'back story' gives authenticity to marketing
- Product quality reflected in new prize-winning expressions
- Owners are able to commit significant marketing funds and distribution strength
- Striking packaging has high shelf appeal

BOX 6.5 CASE STUDY: THE MACALLAN SINGLE MALT

The breakthrough of The Macallan single malt may be dated with uncanny precision. Although this brand, then independent, had featured in Aeneas MacDonald's list of 12 whiskies *"representing Highland whisky at its most distinguished"* (Aeneas MacDonald, *Whisky*, Canongate Books, Edinburgh, 1930) and the company had begun actively selling The Macallan from the early 1960s, it was not until a unknown panel member tasting for the *Harrods Book of Whisky* in 1978 described it as *"a Rolls Royce among malts"* that it began to draw popular attention.

That description, so memorable and so easily reproduced, was repeated countless times in advertising, promotional material and in media articles. At once, it set the image of The Macallan effortlessly above all its competitors. But, hitherto, the distillery had always devoted a considerable amount of its production to the blending market, where the spirit was highly regarded: the rapid growth in sales necessitated a substantial expansion in 1965/1966 when seven stills were added to the existing five and again following the partial flotation of the company in 1968. By 1975, The Macallan had 21 stills in use when a general downturn in orders for fillings persuaded the company to give even greater focus to sales of single malt.

The first marketing director was appointed in 1978 and, in April 1980, an unusual and witty 'teaser' press advertising campaign was begun in the UK. *Harpers Wine & Spirit Gazette* (Feb 1980) commented on the *"slight 'cult' tendency which surrounds the malt whisky drinker"* and thought that The Macallan's strategy of vintage dated bottlings *"could be collector's items in the making"*.

The small black and white cartoon adverts were highly successful and the campaign ran for a number of years, eventually evolving into large 48 sheet poster executions and, where permitted, TV commercials (especially Italy, where the brand had built an early following). In April 1980, Allan Shiach, a member of the controlling family, became Chairman and Chief Executive, strongly influencing the creative and imaginative use of PR, small space advertising (including, controversially for the period, *Private Eye*) and extensive product placement in films and TV through his other career as a noted script-writer. By 1984, despite premium pricing, The Macallan was the 5th ranked single malt in the world; today it lies 3rd.

In 1987, the first 60 Year Old Macallan was auctioned and it fetched the then staggering sum of £5500; the artist, Peter Blake, was commissioned to create the label design for a further 12 such bottles. Subsequently, a number of other limited editions have been created and this remains a key part of the brand strategy. In July 1996, Highland Distillers mounted a successful hostile take-over bid for the company. Later that year, a bottle from the 60 Year Old limited edition sold for £12,000 and, in 1999, the company received its 4th Queen's Award for Exports.

Since the takeover, the brand has emphasised luxury marketing, continuing the release of collectable, super-premium expressions, including the Masters of Photography Series and The Macallan Cire Perdue collaboration with Lalique, setting the world record price for whisky at auction (in 2010). The controversial launch of the Fine Oak range, very different from the traditional sherry cask style, has proved highly successful. Much is made by the company of the so-called 'six pillars', which form the basis of all marketing activity – although they have in fact evolved considerably over the years.

Recently the non-aged 1824 Series has been launched, featuring whiskies whose character is determined by the natural colours from the casks. The impact of this has yet to be seen. Key success factors include:

- Long term focus, both pre- and post-Highland Distillers' ownership
- An early entrant into the single malt category with distinctive, memorable and appealing advertising, strong PR relationships and clear product benefit
- An unusual and distinctive visitor centre and tour
- Owners are able to commit significant marketing funds and distribution strength
- A culture of innovation and well-thought leadership, including successful line extensions to combat stock shortages
- Deep understanding of the luxury goods market

The idea has far outgrown its Scottish roots and today the leading facilities are located as far afield as Taiwan, where the Kavalan distillery receives in excess of one million visitors annually, and in Kentucky where *Whisky Magazine* awarded its 'Visitor Attraction of the Year' for 2010 and 2012 to Buffalo Trace and Jack Daniel's, respectively. Such is the success of the visitor centre concept that, today, it is hard to imagine a new distillery would be constructed without some form of visitor access (Diageo's Roseisle plant is a notable exception, yet there is little doubt that it would attract substantial numbers were it to be opened to the public).

Initially, such tours were free to enter. However, the free entry model has, with a few notable exceptions, been largely abandoned in favour of an entry charge, even if this is subsequently wholly or partly redeemable against shop purchases. Two factors influenced the introduction of charged entry: firstly, distillers realised that coach operators were tending to use a stop at free entry distillery sites to provide toilet facilities for their passengers and that, often, the audience thus delivered was highly unlikely to match the brand's target audience and secondly, because it was then recognised that a modest charge for entry provided a disincentive to the entirely casual visitor allowing the centre's staff better to engage with smaller numbers of prospective customers.

Subsequently, a model has developed of a range of charges. This has recognised both the increasing capital and operating costs of the more sophisticated facilities and the level of knowledge and engagement of the connoisseur consumer. Many consumers, especially single malt enthusiasts, will have made a number of distillery visits, read widely on the subject and hold strong opinions on the subject of whisky. They are increasingly dissatisfied with the standard tour model, often delivered by an inexperienced, part-time or seasonal guide working from a script or by a bland, if visually attractive, video.

Accordingly, a number of premium-priced tour products have been developed. These will be of extended duration, generally with a smaller group of like-minded visitors, possibly gain access to areas not seen by the standard tour, involve a more extended tasting session with rarer and more expensive whiskies and be led, if not by an actual distiller, by a more experienced senior guide, able to deal authoritatively with a wide range of quite probing questions. Prices for such premium tour products can be as high as

£75 (*e.g.*, the Magnus Eunson Tour at Highland Park distillery,
Orkney where a standard tour is available for £6) or even the £150
Century of Whisky Tour at Glengoyne distillery, near Glasgow
(standard tour £7.50). Visitors on such extended tours may spend
as long as 5 hours at the distillery, representing a considerable
investment of their time (let alone money) and a thorough im-
mersion in the brand and product.

There are two principal benefits to be derived from the visitor
centre. Firstly, and not insignificantly, there are direct sales of
whisky and other merchandise. Whisky sales, often of premium or
exclusive expressions, are highly profitable as the distiller is able to
take not only their normal margin as producer, but also all the
distribution, wholesale and retail margins as well. For a small
distiller, or a new start-up, such sales are extremely important. For
any scale of owner, however, the fact that the visitor is effectively
paying to receive marketing messages and to buy the product is of
considerable appeal.

Second, however, is the emotional relationship generated by the
visit (and the consequential purchase) that such sales represent and
the experience of the distillery and brand that the visitor, whether
opinion former, trade or consumer, takes away with them (hence
'experiential' marketing). An effort will generally be made to capture data on the visitor, if no more than an email address, which
can be added to a database for future relationship marketing. The
marketing goal implicit in the visitor centre or brand home is to
build a relationship with the individual visitor that progresses from
casual visitor to supporter to (unpaid) brand advocate, as illustrated in Figure 6.6.

At its best, the successful and well-managed brand home can
promote emotional links with a number of audiences, be they
opinion-forming bloggers, journalists or writers, trade guests or
consumers, moving them beyond functional product delivery to an
emotional engagement with the brand.

6.7   SOCIAL MEDIA AND THE INTERNET

No discussion of this topic would be complete without some reference to the enormous impact of the internet in general and social
media, in particular ,on consumer marketing strategies. In an

The Brand Experience: relationship model

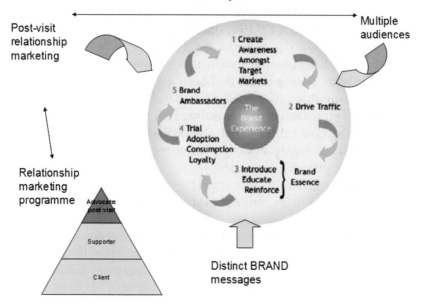

Post-visit relationship marketing

Multiple audiences

Relationship marketing programme

1 Create Awareness Amongst Target Markets

5 Brand Ambassadors

The Brand Experience

2 Drive Traffic

4 Trial Adoption Consumption Loyalty

3 Introduce Educate Reinforce } Brand Essence

Advocate post visit

Supporter

Client

Distinct BRAND messages

Figure 6.6   The Brand Experience Relationship Model seeks to illustrate the key principles of experiential marketing.

extraordinarily short period of time it has become *de rigueur* for almost all brands to have not only a web site but a strategy specifically for the internet as a whole, with particular emphasis on social media.

Remarkably, only ten years ago, Grant E. Gordon devoted less than half a page to 'internet marketing' in one key text,[8] noting that *"the brand needs to have an effective home page and site"* but that *"many of these sites have low visit rates"*. He did, however, identify the opportunity for direct marketing by *"collecting email addresses of interested consumers and entering into a dialogue with them"*.

It is remarkable to note how much changed in that decade, not least the explosion of social media, user-generated content, blogs and internet retailing, all of which have had a dramatic influence on the whisky industry. Indeed, in a very recent development, crowd funding sites are being used to bankroll the creation of new brands of whisky and even provide equity financing for new micro-distilling operations.

The power of social media was dramatically demonstrated in February 2013 when Maker's Mark, a well-known brand of bourbon, announced that, due to very high levels of demand, it was reducing the strength of the product from 90% proof (45% abv) to 84% proof (42% abv). A cut of just 3% abv may not seem hugely significant and the company went to some lengths to explain that their own taste testing panel had been unable to detect the difference and to outline the reasons for the change. Despite this, the result was a storm of consumer outrage expressed in many forceful emails to the company's Chief Operating Officer, Rob Samuels, and a burst of highly negative comment on Twitter and other social media sites. This, in turn, led to a considerable volume of commentary – generally either directly negative or poking fun at the company – in conventional media such that the story became self-sustaining.

Within two weeks Maker's Mark had reversed their decision and resumed shipping supplies at the old, higher strength. This was accompanied by a self-effacing public *mea culpa* from Samuels who wrote: "*You spoke. We listened. And we're sincerely sorry we let you down*".

Such a dramatic reversal of an otherwise apparently commercially rational decision is unprecedented – but may well mark a significant moment in consumer power. The full implications of the Maker's Mark case are yet to be worked out but it is clear that, in the future, marketers will have to take even greater account of social media in their brand communications. In the ultimate irony, it would appear that unopened bottles of the reduced strength Maker's Mark have now attained collectable status, thus ensuring that they will never be drunk!

Whisky has also gathered an enthusiastic and voluble community of bloggers. While it would be fair to say that they vary in expertise, commitment, knowledge and writing ability, it is also fair to observe that the best are considered to be influential and are treated with increasing respect and engagement by the industry. Indeed, most notably in the marketing of single malt whiskies and boutique craft whiskies, considerable PR efforts are now devoted to building relationships with key bloggers who receive regular samples, distillery trips and even dedicated events in an effort to secure favourable comments and reviews for the brand concerned. A few bloggers – most notably the team at CaskStrength.net[9] – have made successful efforts to monetize their site, while others have sought to use their blogging activity to provide an entry into a career within the industry.

While there has been a proliferation of blogs and similar activity on social media sites, this does not appear to have affected conventional publishing. Indeed, today, almost every major whisky market has at least one consumer magazine title dedicated to whisky and a remarkable number of books continue to be published to meet what appears to be a growing demand for knowledge about whisky. The large and increasing number of tasting events also bears testimony to the consumer's taste for more information, combined with an agreeable social event.

6.8 REGULATORY AND CONTROL ISSUES

A fundamental principle to be borne in mind at all times is that any and all marketing should be **legal, decent, honest** and **truthful**. If this simple test is applied to proposed activity at an early stage in its development, significant problems of compliance should simply never occur.

However, beyond that, the advertising and promotion of all forms of beverage alcohol is one of the most strictly regulated forms of marketing. As a general rule, the marketer should assume that there are regulations covering their proposed activity in any market – and the rule is to check first before incurring the reputational damage and possible penalties involved in breaching the relevant codes or legislation.

Responsible marketing of all alcohol involves restraint and maturity on the part of all concerned. The temporary competitive advantage to be gained from deliberately breaking the rules needs to be set against the long-term damage this may – surely, will – cause to the brand and industry, let alone the potential societal damage that could ensue. Any responsibly managed company in the business of marketing alcohol will take steps to ensure that their activity conforms not just to the letter but to the spirit of the prevailing legislation. Additionally, it is incumbent on each individual marketer to monitor their own contribution and look to their professional conscience to guarantee compliance.

This is simply good practice and in the long-term interest of the alcohol industry and society at large. In many markets there are well-developed and highly active advocacy and special interest groups opposed to alcohol marketing who will quickly identify and loudly criticise breaches of local codes, often using specific individual examples to stigmatise the industry as a whole.

Having noted that virtually all markets have strict regulation in place, it is also worth observing that these are not internationally consistent and do vary from market to market and are liable to change over time. What was compliant yesterday may not conform to the standard tomorrow.

The commentary below summarises the situation, so far as the UK is concerned, and is correct at the time of writing (April 2013). The reader is advised to familiarise themselves with the current position in the market of interest before undertaking any activity, especially that directed at the consumer.

The following bodies are involved in the regulation of advertising and promotion of alcohol and specifically whisky in the UK:

- The **Committees of Advertising Practice** (CAP) write and maintain the UK Advertising Codes, which are administered by the **Advertising Standards Authority** (ASA)
- The **Portman Group** is the industry-funded social responsibility body for alcohol producers. They operate a strict code of practice to ensure alcohol is marketed responsibly and not in a way that might appeal to children. This code applies to all pre-packaged alcohol sold or marketed in the UK
- The **Scotch Whisky Association** is the principal industry association for distillers of Scotch whisky and has developed its own code of practice setting out minimum standards for the marketing and promotion of Scotch whisky brands in the UK. It is also the default code for SWA member companies in those export markets where stricter national codes do not exist
- In addition, the retail sale of alcohol is closely regulated and the **alcohol licensing laws** of the UK regulate and control the sale and consumption of alcohol, with separate legislation for England and Wales, Northern Ireland and Scotland being passed, as necessary, by the UK Parliament, the Northern Ireland Assembly and the Scottish Parliament, respectively

All the above maintain websites, where the relevant code of practice may be obtained, and marketers and their agencies and suppliers should familiarise themselves with the details. To summarise, however, the following is extracted from the ASA website:

- The rules are based upon evidence that points to a link between alcohol advertising and people's awareness and attitudes to drinking
- The stringent rules, which apply across all media and are mandatory, place a particular emphasis on protecting young people; alcohol ads must not be directed at people under 18 or contain anything that is likely to appeal to them by reflecting youth culture or by linking alcohol with irresponsible behaviour, social success or sexual attractiveness
- The TV and radio advertising rules contain strict controls about the placement and content of alcohol advertising. Alcohol ads are banned from appearing in and around programmes commissioned for, or principally targeted at, audiences below the age of 18, as well as programmes likely to appeal particularly to audiences below the age of 18

EU law outlined in the 'TV without Frontiers Directive' has subsequently been expanded to cover new media formats, such as digital television. Now called the 'Audiovisual Media Services Directive', the provisions regarding restrictions on alcohol advertising are broadly similar to those of the ASA for the UK. Stricter regulations on alcohol marketing apply to France and countries, such as the Ukraine, Kenya, India and Norway, have banned all alcohol advertising on television and outdoor sites. The importance of checking local regulations prior to beginning any marketing campaign for whisky cannot, therefore, be over-stressed.

With regard to brand websites the precautionary principle has generally been applied, with the majority of sites requiring the visitor to self-certify that they are over the legal drinking age for consuming alcohol in their country. The value of such a box-ticking exercise may be questioned as it is easily subverted but the industry feeling appears to be that at least something has been done to erect a barrier to under-age visitors. Similar steps are taken by many brands on social media sites – a full and satisfactory consensus and common practice is yet to emerge however.

Within the UK, there are a number of powerful and well-funded groups lobbying to reduce the promotion (and consumption) of all alcohol. These include such organisations as Alcohol Concern, British Association for the Study of the Liver, British Liver Trust, British Medical Association, Institute of Alcohol Studies and the Royal College of Physicians. Within England, the Department of Health promotes a Public Health Responsibility Deal, which includes alcohol. The Scottish Government has energetically promoted a policy of minimum pricing on alcohol, currently subject to challenge.

While the topic of alcohol and health is beyond the scope of this chapter, the reader is strongly recommended to visit the websites of the groups listed above (and others) and become familiar with the arguments in this increasingly complex debate. At the same time, it is worth looking at the various responsible drinking campaigns funded at an industry and brand level, which form part of the alcohol industry's response to the prohibitionist tendency. Responsible behaviour is key to the industry's long-term success, prosperity and freedom to operate and everyone working in marketing has a critical part to play in this.

REFERENCES

1. P. Kotler, *Marketing Management: Millennium Edition*, Prentice Hall, New Jersey, 1999.
2. D. E. Schullz, S. I. Tannenbaum, R. F. Lauterborn, *Integrated Marketing Communications*, NTC Business Books, Lincolnwood, Chicago, 1993.
3. W. Meier-Augenstein, H. F. Kemp and S. M. L. Hardie, Detection of counterfeit Scotch whisky by 2H and 18O stable isotope analysis, *Food Chemistry*, 2012, **133**, 1070–1074.
4. A. C. McIntyre, M. L. Bilyk, A. Nordon, G. Colquhoun and D. Littlejohn, Detection of counterfeit Scotch whisky samples using mid-infrared spectrometry with an attenuated total reflectance probe incorporating polycrystalline silver halide fibres, *Anal. Chim. Acta*, 2011, **690**, 228.
5. P. C. Ashok, B. B. Praveen and K. Dholakia, Near infrared spectroscopic analysis of single malt Scotch whisky on an optofluidic chip, *Opt. Express*, 2011, **19**, 22982–22992.

6. See Rochdale Online, *Warning over counterfeit vodka and whisky*, 2012, available online at http://www.rochdaleonline.co.uk/news-features/2/news-headlines/70781/warning-over-counterfeit-vodka-and-whisky (accessed 31st July 2013).
7. J. Pine and J. Gilmore, *The Experience Economy*, Harvard Business School Press, Boston, 1999.
8. G. E. Gordon, in *Marketing Scotch Whisky* in *Whisky Technology, Production and Marketing*, ed. I. Russell, Academic Press, London, 2003.
9. Cask Strength Creative, 2012, blog available online at http://caskstrength.blogspot.co.uk/ (accessed 31st July 2013).

CHAPTER 7

New Whisky Countries

As we have seen, all distilling was originally conducted on a small scale in monasteries, farms and in private homes before a combination of legislation, consolidation of ownership and mechanisation of the process itself concentrated most production into large scale units. However, anecdotally, illicit production of 'moonshine' remains vibrant – indeed, if the availability of stills and other equipment on various well-known internet auction sites is any guide, it continues to flourish. The basic science of the distilling process has long been well understood and it is not intrinsically difficult to make spirit, albeit of varying and inconsistent quality.

A number of guidebooks to home and small-scale distilling have been published, further reflecting the growing interest, and presumably activity, in this area and facsimile reprints of classic works are widely available in the print-on-demand format.

But, until very recently, the trend had been for the production of whisky to be consolidated into fewer and fewer hands and the received wisdom was that whisky could only be satisfactorily made in a few countries. By contrast, small scale distilling of fruit spirits continued in many European countries where legislation, perhaps guided by long-standing traditions and cultural considerations, favoured the artisanal operator. Even in France today much Armagnac continues to be produced by peripatetic mobile stills (Figure 7.1) that

The Science and Commerce of Whisky
By Ian Buxton and Paul S. Hughes
© Buxton & Hughes, 2014
Published by the Royal Society of Chemistry, www.rsc.org

Figure 7.1 A traditional mobile still used for whisky production at the Belgian Owl Distillery, Grâce Hollogne near Liege.

travel to the vineyard when the new wine is deemed ready for distillation, thus avoiding the necessity for capital expenditure on equipment that would lie idle for much of the year.

From the early 1980s, those few small scale whisky distilleries that did continue in operation, such as Royal Lochnagar, Glenturret and Edradour, remained largely as curiosities, operated by their international owners more as tourist attractions than for any intrinsic merit of the spirit. Only Campbeltown's Springbank distillery (a quixotic and eccentric survivor) carried resolutely on; although, in 1987, the well-known whisky writer, Michael Jackson, noted that *"this very traditional plant has not produced for some years"*.[1]

Production of less than a minimum of one million original litres of alcohol (ola) annually was increasingly regarded as anachronistic and uneconomical which, for an industry largely focused on blends, was an entirely understandable stance. A considerable number of smaller plants were closed in various rounds of rationalisation as a result of their cost of production (*e.g.*, Glenglassaugh, Speyburn, Glen Mhor) or because of the cost of adapting them to meet environmental standards (*e.g.*, Rosebank). Ironically, of course, a number of these distilleries have subsequently found new life under new owners and can flourish in today's market.

Prior to the 1990s, even the few producers of single malts who were active released a very limited range of ages and were concerned primarily with building a brand personality in which consistency of taste was a desirable quality. Individual cask variants, finishes and extensive age ranges were largely unknown. This author recalls being firmly told in 1991 by the then managing director of the firm marketing a well-known single malt that the Scotch Malt Whisky Society (SMWS, at that time a callow, irreverent and noisy youth of a society) was a dangerous and undesirable distraction that would likely harm the development of single malt sales by confusing the consumer and encouraging promiscuous buying across the category at the expense of brand loyalty.

While the latter consequence may indeed have followed, the effect of the rapid proliferation of releases appears to have been to feed growing enthusiasm amongst a group of self-selecting connoisseurs with an apparently insatiable appetite for experimentation and novelty. The craft distillers draw much of their support from this group.

However, in the same year that this author received these portentous warnings, Glenmorangie released what was then the first single cask, cask strength bottling from a named distillery (thus aping the SMWS' unbranded releases). The Native Ross – Shire Glenmorangie (Figure 7.2), as it was known, did not survive long in the market but proved a precursor to today's uncountable flood of similar expressions. At much the same time, Glenfiddich's David Stewart had been experimenting with the technique of 'finishing', first explored in The Balvenie Classic and later made explicit with the release of The Balvenie DoubleWood 12 Year Old in 1993.

The first distillery visitor centres in Scotland were opened by Glenfiddich (in 1969) and Glenfarclas (in 1973). Their rivals looked askance at the heretical idea of letting the public inside the distillery itself and, for some time, alternately either ignored or scoffed at

Figure 7.2 Glenmorangie's Native Ross – Shire was the first branded single cask, cask strength bottling.

such radicalism, which was surely destined to fail. However, after some time, this initiative was energetically copied and indeed improved upon, with the result that today the majority of single malt distilleries in Scotland (and many in other producing nations) have sophisticated visitor facilities, or 'brand experiences' as they are often styled, largely operated as a marketing function.

Thus, by the late 1980s, the evident success of the early centres and the growing interest in single malt (then pioneered by Glenfiddich, The Macallan and Glenmorangie) had resulted in a rise in both the number of distilleries with some form of public access and increased availability of single malt expressions. In a perceptive essay,[2] Charles MacLean has shown how the range of Scotch single malt whisky exploded from a mere nine listed by George Saintsbury[3] in 1920 to over 200 identified in the 1989 edition of Jackson's *World Guide to Whisky*.[4]

Since then, the available bottlings have mushroomed to the point where it is now all but impracticable to track every new offering. Indeed, in the 2012 edition of the *Malt Whisky Yearbook*,[5] the editor began the review of expressions launched during the previous 12 months by stating "*it is virtually impossible to list all new bottlings during a year, there are simply too many...*" and apologetically offered a selection of 500!

Historically, while visitors to distilleries were generally encouraged to believe that single malt whisky was a uniquely Scottish product that could only be produced with success in Scotland (a view encouraged by some commentators) such transparency and dynamic growth attracted first curiosity and then an almost inexhaustible thirst for knowledge. The more the distillers shared their expertise with enthusiasts the more the enthusiasts demanded to know – and, in an unexpected consequence of such openness, began to question the orthodox view of an unchallengeable Scottish hegemony.

Thus, the exponential growth in the availability of, for the most part, single malt whiskies has been driven by an enthusiastic consumer group seeking what they believe to be 'authentic' products, associated with values, such as heritage, provenance and artisanal production.

Such factors have been behind the successful re-opening and long-term commercial revival of distilleries, such as Bruichladdich on Islay (Figure 7.3), and copied successfully by the Bourbon industry with the concept of 'small batch' bourbon, essentially a reaction to single malt's international success.

Marketers have been quick to react to this trend and brand positioning aimed at the buyer of premium expressions has rapidly reflected this, especially with increased use of the distillery itself as a 'brand home' (with associated and highly-profitable retail outlet attached), the creation of the role of Brand Ambassador and the promotion of some distillery managers, previously a largely anonymous and unsung group, as distilling superstars with well-developed personalities. This has been greatly assisted by the explosion of public events, such as Whisky Live, WhiskyFest, and many similar tasting occasions, as well as the various specialist whisky retailers that have sprung up serving the consumer through a 'clicks and mortar' strategy.

Armed with this education and a growing confidence, international visitors began to ask whether whisky could be distilled in

Figure 7.3 Distilleries, such as Bruichladdich on Islay, have benefited from consumers seeking authentic craft production and a strongly defined provenance.

their home country and, if not, why not? A few pioneers began experiments and so the phenomenon of artisanal 'world' whisky as we presently understand it was born. However, before considering this phenomenon in greater detail mention should be made of the larger whisky distilleries located outside of the 'classical' whisky producing countries,[†] such as the giant Destilerías y Crianza del Whisky (Whisky DYC) distillery in Segovia, Spain, which began producing whisky in 1962.

In addition to Whisky DYC, today owned by Beam Inc. of the USA, there exist significant whisky distilling operations in Turkey (Mey Icki, est. 1963), South Africa (James Sedgwick, 1990), Amrut, McDowell's and John in India (1948, 1988 and 2009, respectively) and the remarkable Muree Brewery in Rawalpindi, Pakistan, where whisky has been distilled since 1860. More recently, the substantial Kavalan distillery has been built in Yuanshan, Taiwan, with the potential to produce around 2.5m ola per annum. There is, of course, a very large industry in India producing Indian whisky

---

[†]For the purposes of this commentary, the 'classical' whisky nations are considered to be Canada, Ireland, Japan, Scotland and the USA.

for the domestic market; that product, however, does not meet international standards for whisky and is therefore not considered here.

In at least one case, that of Mackmyra in Sweden, what began life as a small craft distilling operation has been substantially expanded. Established as recently as 1999, Mackmyra was inspired by visits to Scotland made by the original founders but has subsequently evolved a distinctively Swedish style and identity. The original distillery had a capacity of 170 000 ola but, by 2011, in the first phase of a £50m, ten year plan, an entirely new distillery, of radical appearance, was built at Gävle, together with warehousing and visitor facilities. This new distillery virtually trebles Mackmyra's potential output to ca. 500 000 ola but the long-term plan is for production to reach 4m ola annually. Such a level of output would make Mackmyra a very significant producer of single malt whisky (Figure 7.4).

The level of investment required to fund a dramatic expansion such as this, especially as the combined output is destined for sale as a single malt, demonstrates the ambition of at least some parts of the craft distillery sector and the market's ready acceptance of a product from a non-traditional producing nation. Following in Mackmyra's footsteps, a further five small distilleries are now producing whisky in Sweden.

Turning then to small-scale (*i.e.*, less than 100 000 ola pa capacity) 'craft' producers, amongst the earliest artisanal pioneers outside the 'classical' regions were Austria's Reisetbauer and Waldviertler Roggenhof distilleries (both 1995), the Belgian Owl distillery (1997), Warenghem (1994) and Distillerie Guillon (1997) in France, the venerable Blaue Maus distillery (Germany, 1980) and Locher in Switzerland (1998); as well as Lark (1992), Bakery Hill (1998) and Tasmania (1996), all in Australia, and Zuidam in Holland (1998). Since the end of the 1990s the number has greatly increased and whisky is now produced in Australia, Austria, Belgium, the Czech Republic, Denmark, England, Finland, France, Germany, Liechtenstein, The Netherlands, Pakistan, Russia, South Africa, Spain, Sweden, Switzerland, Turkey, Wales and, in all probability, elsewhere as well.

The 'blue riband' for being the first of these innovators would thus appear to be held by the Fleischmann family of the Blaue Maus distillery in Germany. Bill Lark of the eponymous distillery

Figure 7.4 Mackmyra's new distillery illustrates the ambition of some new world distillers. Photo credit: Johan Olsson.

in Tasmania is generally considered the 'father' of a now vibrant Australian distilling movement and, excluding Edradour, Anthony Wills of Kilchoman on Islay has inspired a renaissance in Scotland and elsewhere.

In addition to the above, small scale distilleries have also appeared in Canada, Ireland, Japan, Scotland and the USA. By far the most vibrant micro-distilling sector today is that in the USA, where many sites produce primarily gin or vodka, largely because of the immediate cash-flow generated by these products. However, most either currently produce some version of whiskey or aspire to do so. A number have produced highly experimental whiskies, sometimes referred to as 'weird whiskies'.

According to the American Distilling Institute (ADI) at the date of its formation in 2003, there were 69 craft distillers holding licences from the Alcohol and Tobacco Tax and Trade Bureau (TTB). Currently, the ADI estimates there are over 350 US craft producers, with another 50 or so under construction. Partly driven by whisky's increased fashionability and partly by a desire for self-employment (not to say self-fulfilment) following the economic crisis, *legal* small scale distilling has perhaps never been as popular in the USA as it is today. While there were undoubtedly many more stills operating around the time of the Whiskey Rebellion (1791 – 1794) and there were, of course, uncounted numbers of small bootleg distillers operating during the prohibition period, these were of necessity parochial and did not enjoy the fashionable cachet currently attached to craft distillation.

The movement is less well developed in Canada, Ireland and Japan, although it does feature in all; however, it has begun to gather momentum in Scotland, based on the perceived success of Edradour (est. ca. 1837, originally as a farmers' co-operative) and Kilchoman (2005, Box 7.1). There are sites currently operating at Daftmill in Fife, Abhainn Dearg on the island of Lewis and the 'illicit sized' still at Drumchork Lodge Hotel, Aultbea (the Loch Ewe Distillery). Further developments are under construction at Annandale in the Borders region, Lochaber in Ardnamurchan, Kingsbarns in Fife and potentially at a number of further sites.

Despite a number of high profile failures, which might have been thought likely to dampen entrepreneurial enthusiasm and small investor interest, further projects for Scotland are in the fundraising phase. Boutique distilleries are also planned for England in London and the Lake District to join those currently operating in Cornwall, Suffolk and Norfolk.

Such is the interest in the UK that a Craft Distillers Alliance (CDA) has been proposed;although, at the time of writing, this has yet to take final shape or commence significant operations. The proponents of the CDA maintain that the interests of small-scale distillers are markedly different from those of the industry giants represented by the Scotch Whisky Association (SWA) and that a specialist lobbying and advocacy group is therefore required. It is proposed that membership of the CDA will be restricted to those distilleries producing no more than 140 000 ola annually.

BOX 7.1 CASE STUDY: KILCHOMAN DISTILLERY, ISLAY

The Kilchoman distillery is located on the west coast of Islay and, as it proudly boasts, was in 2005 the first distillery to be opened there for 124 years.

Styled as a farmhouse distillery, output at full production is around 120 000 ola per annum (2012 target). The original intention was to keep all the production at the distillery, from barley grown on the surrounding Rockside Farm, through to malting and distilling to warehousing (Figure 7.5). In fact, only around one third of the annual production is malted on site from locally-grown barley, with the remainder coming from the Port Ellen maltings. The spirit is kept strictly separate, however, and 100% Islay production sold at a premium.

Although now operating successfully, Kilchoman had a difficult gestation and birth. There were early management challenges, especially in obtaining the appropriate advice and technical guidance; limitations in the design meant that it was difficult to achieve sufficient steam pressure to satisfactorily operate the stills; there was a fire in the maltings and the original cost projections proved unrealistic. It was necessary to raise significant additional funds to complete the project and ownership has, therefore, been diluted for the founding team. The design and size of the still house mean that future expansion is problematic and there is limited warehousing capacity on site.

Notwithstanding the above, the products have been enthusiastically welcomed by single malt enthusiasts. The location on Islay, and the distillery's unique scale, ensure a steady flow of visitors to the shop and café, which provide vital extra income. The Islay connection and Kilchoman's strong brand proposition and niche marketing enables the products to command premium prices for what is still young whisky. Exports have been successfully achieved, despite the inevitable difficulties (not unique to Kilchoman) in finding distributors. Marketing has been aided by positive comments from whisky writers and bloggers and the web provides vital market access and the ability to reach consumers directly.

The stills are amongst the smallest in Scotland. The wash still receives a 3000 L charge and the spirit still 1600 L. Currently, around 22 casks are filled weekly, using first fill bourbon barrels from Buffalo Trace Distillery, Kentucky, USA and first fill oloroso sherry butts and hogsheads from Miguel Martin, Jerez, Spain. Distilling is carried out under the manager, John McLellan (formerly the manager at Bunnahabhain), while founder and MD, Anthony Wills, remains heavily involved in sales and marketing activities, as well as general management.

After a challenging beginning, the company is now profitable and fulfilling the vision of the original founders.

Key success factors include:

- Experienced distilling management, while ownership had relevant sales and marketing skills
- Supportive, well-funded shareholders provided additional equity
- Location and reputation of Islay as distilling centre
- 'Cult' following for Islay malts generated positive PR from early stages
- High-margin income from visitor operations – retail and café sales
- Distinctive marketing and good packaging justifies premium pricing

The CDA appears to be loosely modelled on the ADI and it is apparent from the proliferation of craft distilleries in the USA that the existence of an active and energetic industry association can encourage the sharing of knowledge and promote favourable legislation.

The ADI, for example, have been active in campaigning to help support the recent expansion of craft distilling and to encourage further growth of small USA-based spirits manufacturing by creating a concessionary Federal excise tax rate for small-scale distilled spirit makers that mirrors the current reduced tiers for small beer and wine producers. The ADI also publishes or sells a range of educational books, promotes an active on-line discussion forum and organises an annual conference and an eagerly-contested awards competition.

Figure 7.5   The floor maltings at Kilchoman on Islay. See the case study in Box 1 for additional details.

A particularly significant development for the future of micro-distilling in the UK has been recent clarification of the law on still sizes permitted by HM Revenue & Customs (HMRC, formerly HM Customs & Excise). For many years it had been the established and understood position that the minimum acceptable still size was 400 gallons (1818.44 L, or 18.2 hL), the reasoning being that any still smaller than this would be too easy to conceal, remove and transport thus facilitating illegal distilling. The UK legislation had been consolidated over many years, dating back certainly to the 1823 Excise Act and extensively revised and amended since then. Custom and practice (and quite possibly the convenience of HMRC) had combined to create a situation in which the 1800 L limit was taken as fixed and given by all concerned. An industry increasingly controlled by multi-national corporations and concerned largely with large-scale production for blending saw no reason to challenge the *status quo*; indeed, it might be said, there was benefit from the perceived barrier to entry presented by this size limit.

This was first challenged in 2002 by Frances Oates and John Clothworthy, owners of the Drumchork Lodge Hotel in Aultbea, who wished to offer their guests the experience of distilling on a very small pot still in an imitation of a 'traditional' manner 'using illicit methods'. After a two year campaign they were able to show that certain key conditions of the 1786 Wash Act relating to Highland Scotland had never been repealed and accordingly received permission to distil in a 120 L pot still − after which the loophole was promptly closed. Their Loch Ewe distillery could thus claim a unique status as unarguably the smallest licensed distillery in Scotland.

There, matters appear to have rested and subsequent operations (*e.g.*, Kilchoman) were developed in the belief that the 1800 L rule was absolute and could not be challenged. It has recently (July 2012) transpired that this is not, in fact, the case.

A close reading of the relevant legislation by Alan Powell, an independent excise duties consultant but formerly an HMRC headquarters official, who had formulated much of the relevant policy while employed by HMRC, revealed that there was no such bar. Powell had been engaged to advise the management of the proposed London Distillery Company on their application for a distilling licence, their goal being to establish an 'artisanal' distillery in the capital.

The Alcoholic Liquor Duties Act, 1979 (ALDA), section 12(1) states: "*No person shall manufacture spirits... unless he holds an excise licence for that purpose...*". ALDA, s12(5) states: "*Where the largest still to be used... is of less than 18 hectolitres capacity, the Commissioners may refuse to grant a licence or may grant it only subject to such conditions as they see fit to impose...*". The default position therefore is that the commissioners of HMRC have the power to issue a distiller's licence upon receipt of an application. HMRC *may* only refuse to issue a licence, or apply conditions to the licence approval, where the largest still is less than 18 hL. The law (consolidated from many years ago) provided HMRC with this discretion because of the possible risk that such a still might be moved from what would then have been the 'entered premises' for unrecorded illicit use. In current times, this risk is negligible and certainly not credible where the distillery is located in a restricted urban area.

HMRC's internal guidance on their official website states: "*...it is not our normal policy to grant approval* (for a licence) *where....*

the largest still used has a capacity of less than 18 hectolitres, or an equivalent throughput could not be achieved in an 8 hour shift using a patent or continuous still. However, we may consider licence applications in respect of stills below 18 hectolitres where there are satisfactory controls in place to protect the revenue and the required control resources are not disproportionate to the amount of revenue involved. All applications to manufacture spirits in stills of less than 18 hectolitres are to be referred to the (HQ) *Alcohol Team for consideration on a case by case basis".*

The legal basis then for granting a distilling licence for any size of still is "*to protect the revenue*" and HMRC are required to act reasonably in applying their judgement. Furthermore, since 1995, it has been the case that HMRC's decisions can be appealed at a First-tier Tax Tribunal, which is a considerably less onerous, time-consuming and expensive procedure than the previous system of judicial review through the courts.

The implications of this policy for small-scale distilling in the UK – not just of whisky, but of all spirits – are potentially enormous. While the aspiring distiller will still require appropriate approval from HMRC under the Customs and Excise Management Act, 1979, section 92, together with a registration under the appropriate regulations for a bonded warehouse, the perceived barrier to entry has been significantly lowered.

Most importantly, the capital cost of establishing a small distillery has been drastically reduced and, in consequence, the next decade may well see a number of new entrants operating stills considerably smaller than 1800 L. Distilling in the UK may well be on the verge of a significant renaissance as a revival of its small-scale roots now appears highly likely, leading to greater diversity in the industry.

There appear, therefore, to be a number of factors which have contributed to the international rise in craft distilling, including:

- Increased availability of a range of single malt expressions, and
- Increased public access to distilleries, leading to
- A growing connoisseur and enthusiast market with access to a greater range of educational resources, such as books, magazines, whisky fairs and festivals, formal and informal training courses (some self-taught) and a proliferation of information available *via* the World Wide Web, combined with

- Pioneering craft distilleries, supported by formal and informal 'associations', subsequently aided by
- An increasingly favourable legislative and regulatory environment, and
- A growing market demand for whiskies that are seen to be 'authentic' or which, by contrast, are seen to be 'innovative' or in some way challenge convention, served by
- The growth of specialist retailers, many with sophisticated online facilities

If such varied factors have combined to stimulate the movement to craft distilling, what are the production implications, quality assurance and control issues associated with it?

A small scale is, of itself, no barrier to quality or consistency. The reputation of Highland single malts was built from the 18th century onwards on the fact that the small Highland still created a product considered greatly superior to the more rapidly distilled Lowland whiskies in their larger stills. These, with their broad, saucer-shaped design could be discharged as often as 94 times in a 24 hour period[‡] but produced spirit which the poet Robert Burns described as *"a most rascally liquor"*.[§] While the reputation of Highland whisky was no doubt enhanced by the illicit thrill of a clandestine product, often smuggled to the end user – as Sir Thomas Dewar later observed in a different context: *"if you forbid a man to do a thing, you will add the joy of piracy and the zest of smuggling to his life"*,[6] – the cachet attached to regions, such as Glenlivet, has lasted to this day.

It was famously Glenlivet whisky that, in 1824, George IV consumed so enthusiastically on his grand visit to Scotland, memorably described by Elizabeth Grant as *"long in wood, long in uncorked bottles, mild as milk, and the true contraband goût in it"*.[7] As a curious aside, while the reference to aging in wood is clear the allusion to *"uncorked bottles"* remains to be satisfactorily explained.

[‡]The vast majority of the spirit produced was designed for rectification by English gin distillers and is unlikely to have been consumed as whisky - until 1788, when the passage of The Lowland Licence Act effectively closed the English market and resulted in a temporary flood of raw spirit onto the Scottish market, forcing a number of distillers, both large and small, out of business.

[§]In a private letter to John Tennant of Glenconner in 1788.

However, from the advent of a more global commercial industry driven by sales of blended whiskies, a principal concern has been to achieve consistency of spirit quality. One method of ensuring this was to increase the scale of the plant. While Stein and Haig of the Lowland distilleries at Kennetpans, Kilbagie and Canonmills had been enormously innovative, greater and greater scientific rigour began to be introduced into whisky production. Early efforts were made by Duncan McGlashan of Edinburgh's Caledonian Distillery who produced his *Table of Proof Spirits* in 1877 and the better-known J. A. Nettleton, author of the magisterial *The Manufacture of Whisky and Plain Spirit*,[8] which was to remain the standard text on the subject for many years.

While it has now been largely airbrushed from whisky's history, there appears little doubt that much whisky was extensively adulterated during the 19th century, when whisky (especially Scotch whisky) enjoyed a dubious reputation as the drink of choice of the lower classes. Pioneering work was undertaken in Glasgow, especially, first, by Dr James St Clair Gray, who in 1872 presented evidence in the *North British Daily Mail* of widespread adulteration of whisky purchased in low-grade pubs and shebeens.[9] Dr Gray died at the age of 27 from diphtheria in 1874.

His work was vigorously disputed by the highly-respected analyst, Robert Tatlock, who later served as Glasgow's city analyst and founded the firm of Tatlock & Thomson, which continues to this day to serve the whisky industry. Notwithstanding his criticisms of Gray, it is now indisputable that substances, such as 'Hamburg sherry', 'prune wine' and 'cocked hat spirit', were to be found in many blenders' and publicans' cellars, a fact made much of by the principal Irish distillers in their campaign for pot still distillation.

Tatlock was later to give evidence on behalf of the grain distilling industry to the 1908 Royal Commission arising from the 'What is Whisky?' case (see chapter 1).

Sir Peter Mackie of White Horse and Little Mill distillery fame established a laboratory in Campbeltown, where S. H. Hastie undertook pioneering work on yields. The self-effacing Hastie later described his endeavours somewhat modestly, merely remarking in a lecture to the Institute of Brewing that "*the application of science... to the control of pot still distillation processes is still in its infancy after a long series of intermittent attempts to make practical use of laboratory work*".

The Distillers Company Ltd (DCL) established their pioneering laboratories at the Menstrie Research Centre, near Stirling, on the site of the old Dolls distillery (later known as Glenochil) shortly after World War II. The first head was Dr Magnus Pyke, of subsequent TV fame, but most of the significant scientific and research work was done under a later head of the laboratory, Dr Bob Duncan. Today, it is Diageo's Brand Technical Centre.

The current lead on co-operative industry research is taken by the Scotch Whisky Research Institute (SWRI). Their low public profile and somewhat bland website reflect the fact that the SWRI exists to support its members, who represent more than 90% of the Scotch whisky industry's production and that the SWRI itself has no consumer role whatsoever.

Its history can be traced back to a predecessor organisation, Pentlands Scotch Whisky Ltd and, even prior to that, the Inveresk laboratories of Arthur D. Little Inc., one of the longest-established global consulting firms.

Arthur D. Little were working with a number of distillers in the 1970s and early 1980s when it was recognised by John McPhail of Robertson & Baxter Ltd (blenders and part of The Edrington Group) that there would be greater strength in a co-operative, industry-owned organisation. Accordingly, the Arthur D. Little interests were acquired by a new company, Pentlands Scotch Whisky, and offices established in Slateford Road, Edinburgh.

Pentlands did much work on flavour-related and spirit yield issues but is best-known, at least to consumers, for their pioneering work on the 'whisky wheel', starting around 1979.

This was the first systematic attempt to define the language of whisky tasting and is now the accepted way of tabulating aromas and flavours. At the time, it was novel. Initially, the wheel was intended for industry use only but, with increasing consumer interest and growing connoisseurship, a more extensive consumer version (Figure 7.6) was developed in the mid-1980s by Charles MacLean, amongst others (for which we are forever in his debt).

Recognising the need to expand Pentland's work, it grew into the Scotch Whisky Research Institute, established in October 1995, and is now located in purpose-built facilities, the construction of which was supported by The Robertson Trust (the charitable trust that is Edrington's ultimate parent).

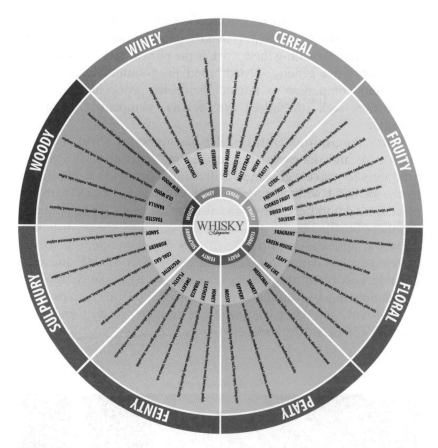

Figure 7.6 A consumer version of the 'whisky wheel'. Courtesy of Whisky Magazine, www.whiskymag.com.

The SWRI's aims are set out on their website and are as follows:

- Maintaining and improving product quality
- Safeguarding product integrity
- Adding value by enhancing the use of raw materials and improving manufacturing processes
- Providing the understanding to facilitate beneficial changes in manufacturing processes

As they say, "*the Institute works closely with Universities, other research facilities and Scotch whisky companies. It is able to carry out joint pre-competitive research on behalf of all its members as well as specific projects for single companies. Its laboratories are UKAS-accredited to ensure the highest quality in all services"*.

Key to understanding their work is the phrase "*pre-competitive*". SWRI work on generic, broad-based programmes and it is for individual companies to pursue their own work independently where they seek (or indeed are seeking) competitive or brand advantage.

Today the aspiring craft distiller has access to a wide body of knowledge and the experience of the first generation of artisanal producers to provide guidance. Still manufacturers have responded to the demand with new designs of small stills, many with sophisticated computer control; the distiller has access to high quality supplies of malted barley at a consistent standard; the contribution of wood and the importance of wood management is better understood; academic training and study is available and it is possible, cost permitting, to engage experienced freelance distilling and engineering consultants. There is every opportunity to produce a consistent, standardised spirit that should age well and meet exacting competitive standards (Figure 7.7).

Figure 7.7   Stillroom at Finland's Teerenpeli distillery, with the stills and spirit safe by Forsyths. See the case study in Box 7.2 for more details.

BOX 7.2   CASE STUDY: TEERENPELI DISTILLERY, LAHTI, FINLAND

Teerenpeli is located in the city of Lahti (population: 100 000), about 60 miles north-east of Helsinki.

The distillery, which may be found in the basement of the restaurant of the same name, was established in 2002 by owner Anssi Pyysing, who previously operated a pub brewery here (this proved very successful and has been expanded to occupy its own purpose-built, stand-alone site).

The plant was supplied by Forsyths of Rothes in Scotland and constructed to what was a then completely unique design, partly based on advice from a Scottish friend of Pyysing's who was, at that time, manager of a large Scotch malt whisky distillery. He also provided some technical assistance and *ad hoc* training in distilling, although Teerenpeli now employ their own distillery team.

A mash tun of 350 kg capacity feeds a single pair of stills, with the wash still holding 1500 L and the spirit still a modest 900 L per charge. At full capacity, Teerenpeli could theoretically produce around 15 000 L of spirit annually. Fortunately for the distillery, Lahti is a major centre for malting and, from the start, the distillery has used locally-sourced Finnish malt, currently peated to a modest 7 ppm of phenols.

Encouraged by the success of his brewing venture and supported by cash flow from this and his small chain of restaurants, Pyysing has been able to source high-quality casks (a variety of types are used) and had no need to rush spirit out to the market until he was satisfied that it met his quality requirements. Indeed, unusually for an operation on this scale, he has been able to retain some casks with the aim of maturing them for up to 20 years.

The distillery has enjoyed good local market support from the outset with a group of enthusiasts willing to purchase individual casks and Alko, the Finnish state alcohol monopoly, listing the product for retail sale. It may also be found in the group's restaurants and in the duty-free area at Helsinki airport. The look of the distinctive packaging was created in Lahti's renowned Institute of Design.

Despite its premium price (currently around the equivalent of
£90 for a full bottle), the distillery regularly sell all of their annual release and are considering plans for a significant expansion on a new site to meet demand. The major challenge now
facing Teerenpeli, apart from finding new markets for the increased production, is warehousing and maturation in the ex-
tremes of the Finnish climate.

Key success factors include:

- Experienced entrepreneurial management team with relevant production and marketing skills from related
  operations
- Positive cash flow from other businessed avoided the requirement for external financing or partners, making for
  quick decisions and clear lines of authority
- Mentorship from an experienced distiller
- Top quality, custom-designed plant from a renowned
  supplier
- Strong brand identity and sense of local/national ownership
- Unique market position as Finland's only whisky distiller

Yet, ironically, while the consumer certainly does not look for
bad or poorly flavoured whisky, consistency, as such, is not especially a quality that is valued in small-scale production. Most, if not
all, of the output is bottled as distinct batches, often single cask,
and the enthusiast market takes enormous pleasure in detecting
and debating the variances between them.

Indeed, more than this, some distillers have made a considerable
virtue of their extensive range of expressions. Among the larger
small producers, Bruichladdich have pursued this as a conscious
and deliberate strategy and even the largest brands offer some
limited edition or short run expressions, sometimes offered exclusively for sale at the distillery. The high profit margin on such
sales and the enthusiastic advocacy of the product by the purchasers to their friends and family make these expressions especially attractive to the brand owner.

Beyond this, there is a trend for highly experimental releases, a
phenomenon especially marked amongst US craft distillers, such as

Balcones and Corsair, and in Switzerland by The Whisky Castle.¶ In a radical departure from conventional practice, whiskies are now being made with ingredients such as wildflower honey, turbinado sugar and figs; a number of alternative cereals have been tried and hopped washes are also being employed (a nod to whisky's close links to beer); smoke has been injected into the casks of maturing spirit and there have been some drastic experiments in methods of cask management to manipulate flavour. The Whisky Castle makes their Snow Whisky using water from melted snow – of necessity, it is only an occasional production and very labour intensive, but the limited supplies sell out quickly. Furthermore, additional staves have been inserted into casks and varying levels of wood char have been explored and, at the Tuthilltown distillery in New York State's Hudson Valley, a technique known as 'sonic aging' is employed, involving exposing the casks to a steady vibration caused by the bass notes of loud rap music!

Some small distillers (*e.g.*, Tuthilltown, USA and St George's in England) have employed old shipping containers as warehouses, claiming that the accelerated aging promoted under such conditions allows them to bring their product to market in less time than would be the case if more traditional storage had been employed. However, while the understanding of maturation has increased, such that excellent whisky is now being produced in climates previously considered inimical to good maturation (as witness the single malt whiskies of Amrut, India and Kavalan, Taiwan), good warehouse temperature and humidity control is far from universal.

Whisky enthusiasts, however, currently appear to embrace such innovation and variance with great glee, passionately debating every nuance of flavour in on-line forums and at whisky fairs and festivals.

The existence and ubiquity of social media has enabled small distillers to communicate directly with their target audience and consumers in a uniquely personal way. Not only has this enabled them to compete effectively with the brands of their giant competitors (with their substantial marketing budgets), it also creates a genuine one-to-one conversation that builds an emotional

---

¶Distillers in Scotland are more constrained by legislation if they wish to style their product 'Scotch whisky'.

relationship that large brands find it hard to emulate. Such trends are discussed in greater detail in chapter 6.

Currently, then, market trends favour the development of the artisanal or craft distilling sector. More new entrants in more countries may be anticipated. However, this sector faces a number of challenges. Apart from the well-documented problems of raising both equity finance for construction and fit-out and securing sufficient working capital, there are challenges in gaining access to markets, both in appointing and managing distributors and in meeting the regulatory requirements of varying countries.

The shortage of capital has led a number of small distillers to purchase the cheapest equipment available, often second-hand or designed for a different purpose (such as fruit stills being employed for whisky production) or, making a virtue of necessity, cannibalising their own design of plant from whatever equipment they can find, with predictably inconsistent results. The tendency of some operators to regard themselves as 'artists' rather than artisans, encouraged by enthusiastic commentary from the writing and blogging community, has introduced further idiosyncrasies into the market.

As more small distillers enter the market, the sector will become more crowded, blurring distinctive individual identities and making it harder to establish and maintain a unique positioning. By definition, there can only be one "*X's smallest distillery*", for example, and for the late entrants the problem of brand positioning becomes ever more acute as the artisanal message secures a reduced competitive advantage and is less of a distinguishing feature.

Moreover, the promiscuous buying behaviour of the consumer is both an opportunity and a threat. The opportunity to sell the first bottle of something new in a market driven at least in part by novelty is clear, but the challenge of selling a second and subsequent bottle to that same buyer is consequently greater and represents a real difficulty for the small distiller, who may lack significant marketing resource but is looking to develop a degree of brand loyalty.

Larger brands have already begun to compete with a range of small batch or single cask offerings, which combine the appeal of novelty with the quality reassurance of a known brand. A number of larger brands, such as Highland Park, have successfully exploited the market for collectible editions and sophisticated marketing campaigns, such as the Ardbeg Committee, which combines scale with apparently home-spun appeal.

Supplies of good quality casks may also represent a problem in the future. A number of very significant expansions in whisky production, especially Scotch whisky, have been announced amounting to more than the whole output of the craft distilling sector combined. Very soon, this whisky will require casks for maturation and the risk to the small distiller must be that the supplying bodegas, American distillers and cooperages will naturally give priority to their long-established, larger and more creditworthy customers over the small and intermittent orders of the craft distilling sector.

Finally, considering the extent to which this market is currently driven by fashion, there remains a risk that a fickle market could simply move on and lose interest in the small distiller. The authors can vividly recall that, as recently as the mid-1980s, the currently enormously fashionable peated style favoured above all by Islay's distilleries and currently *à la mode*, was all but impossible to sell. Hard though it may be to conceive, distilleries such as Port Ellen and Ardbeg were not shut and mothballed and others put onto very low levels of production on a whim. Less than 25 years ago, heavily peated whisky was not heavily demanded by either the blending industry or single malt consumers. Should fashion move on, the position of the craft distiller could rapidly become a perilous one.

Conversely, the small distilling movement does present some challenges to larger brands, perhaps analogous to the impact of the UK's Campaign for Real Ale (CAMRA) on the brewers of keg beers. While, in volume terms, it would seem unlikely that any one craft distiller can make any meaningful impression on the industry's sales, the aggregate effect will be more interesting and it is possible that, as the movement gathers momentum, there will be a measurable effect.

More significant may be the impact measured by value, as the small distillery sector has, to date, been able to achieve surprisingly high levels of premium pricing for young whisky. If this can be sustained (and the pressure of further new entrants and the decay of the novelty factor may militate against this), then the impact will be more profound. However, while this sector has successfully appealed to the whisky enthusiast and collector, it has not been able – and seems unlikely to be able – to achieve the super-premium levels of retail pricing achieved by brands such as The Macallan, Highland Park, Glenfiddich and The Dalmore amongst others.

Perhaps the sector's greatest influence will be to encourage innovation and experimentation, possibly even re-defining what is meant by 'whisky' in some consumers' minds, even if the statutory definitions do not change. While the ever-vigilant SWA tightly polices the Scotch whisky regulations, small distillers internationally are free to disregard these with impunity, making their own styles and expressions.

Such 'whiskies' are unlikely, however, to represent a significant or sustained challenge to established whisky orthodoxy. However, the advent of larger-scale producers in the new whisky countries may well represent a greater risk.

There now appears no technical reason why high-quality whisky cannot be produced in the most unfavourable of climates. While the Scotch and Bourbon whisky industries have invested heavily in the BRIC markets and much new capacity has been installed at great cost to meet the anticipated demand that will follow as their newly-affluent middle classes aspire to purchase imported whiskies, there would not appear to be any significant technical barrier to entry for new producers.

Hypothetically, then, a new producer might emerge in, say, China utilising the latest technology to produce whiskies. This could, in theory, include the use of a column still to produce a 'single malt' (impossible in Scotland) and advanced maturation techniques that would bring fully mature spirit to market in a timetable that would be impossible elsewhere. The Kavalan distillery in Taiwan (Box 7.3 and Figure 7.8) and the various Indian producers of true single malt have already shown that their 3 and 4-year old spirit can compete successfully with Scotch whisky of 10–12 years maturation.

Such a producer would enjoy cost advantages in production and shipping, potential protection within tariff walls and, with skilful marketing, patriotic support from a consumer able to purchase a product that looked and tasted like a high quality import at local market prices. One may thus envisage a scenario in which established producers could be faced with significant competition from new world producers. Why, one might ask, would our hypothetical new market distiller not wish to compete for a share of a growing, profitable and fashionable market on their own doorstep?

Figure 7.8   The impressive Kavalan distillery in Taiwan. New world producers do not lack ambition. See the case study in Box 7.3 for more details.

The Scotch whisky industry was once notoriously complacent. Aeneas MacDonald, writing in 1930 at a time of severe depression in the industry, characterised it as: "*a body of men... assured of their commercial acumen*"[10] even as distilleries closed their doors forever. As late as 1951 the official SWA line was to disparage the upstart Japanese whisky industry as producing a pale imitation of Speyside whisky, which the author found "*not good although drinkable*".[11]

Such an attitude prevailed in Dublin in 1878 when the Thomas Street distillery of George Roe & Co. was the largest pot-still distillery in the world. Along with three fellow distillers of similar scale, Irish whiskey was dominant in world markets. While many factors contributed to the catastrophic decline of Irish distilling complacency, an unwillingness to innovate and embrace the new technology of blending and a blind belief in their traditional practice was to prove commercially disastrous. History suggests, therefore, that the currently well-entrenched and dominant position of the traditional producers is far from impregnable and they would do well to consider potential challenges.

BOX 7.3   CASE STUDY: KAVALAN DISTILLERY, YUANSHAN, TAIWAN

Taiwan is a significant export market for Scotch whisky. According to the SWA, in 2009 it was the 18th largest market by volume but 10th by value, demonstrating its importance for premium brands, which was reinforced by a 44% jump in value in 2010 (the global market increased at roughly half this rate).

The King Car group of companies is a widely-diversified, locally-owned business with significant interests in pharmaceuticals, orchid farming, coffee and soft drink production and coffee retailing. The controlling Lee family were long standing whisky enthusiasts and collectors who, in 2003/2004, were determined to build a distillery that could rival the quality of the best single malts.

The distillery was built on a green field site in Yi-Lan province about 1 hour from the capital Taipei and was producing by 2006. Currently capable of producing around 1.2m ola annually, the design is such that output could be quickly doubled if the demand justified this. The original equipment was supplied by Forsyths of Rothes following visits to Scotland by the Kavalan R & D team and the distillery continues to be advised from Scotland by consultant Dr Jim Swan.

The first releases encountered in the west were tasted blind in January 2010 for *The Times* newspaper by an experienced panel and placed some very young Kavalan spirit against a number of equally youthful samples from Scotland and England. The panel were astounded by the quality of the Kavalan samples and, to general astonishment and some consternation, placed them a comfortable first in this group. Since then, Kavalan have won an enviable number of prestigious international awards, confirming the quality of the whisky, and have been highly commended in some well-known whisky guides.

There is a strong emphasis on quality throughout the production process, evident in the narrow cut from the spirit run, the high filling strength and the stress laid on wood selection and warehouse management, where the regime is designed to offset Taiwan's high summer temperatures.

Kavalan have not yet begun international marketing in earnest but have built visibility through sound PR activity and success in well-regarded competitions. A substantial visitor centre attracts around one million visitors annually, many from nearby China, and it may be that stocks are being built for a serious sales effort there in due course.
Key success factors include:

- Experienced and committed management team and ownership
- High levels of investment available from group funds
- Mentorship from an experienced distilling consultant
- Top quality, custom-designed plant from a renowned supplier
- Strong brand identity and sense of local/national ownership
- Unique market position as Taiwan's only whisky distiller
- Prestigious awards and expert endorsement
- Ability to plan and fund for the long-term

In chapter 6, the importance of marketing and brand development is discussed and this, of course, represents one of the areas of current competitive advantage for the established producers.

REFERENCES

1. M. Jackson, *The World Guide to Whisky*, Dorling Kindersley, London, 1987.
2. C. MacLean in *Beer Hunter, Whisky Chaser*, ed. I. Buxton, Classic Expressions, Pitlochry, 2009.
3. G. Saintsbury, *Notes on a Cellar Book*, Macmillan, London, 1920.
4. M. Jackson, *The World Guide to Whisky*, Dorling Kindersley, London, 1989.
5. The Editor in *Malt Whisky Yearbook*, 2012, ed. I. Ronde, Magdig Media, Shrewsbury.
6. Quoted in *The Wit & Wisdom of Tommy Dewar*, John Dewar & Sons Ltd, Glasgow, 2001.
7. E. Grant, *Memoirs of a Highland Lady*, John Murray, London, 1898.

8. J. A. Nettleton, *The Manufacture of Whisky and Plain Spirit*, G. Cornwall & Sons, Aberdeen, 1913.
9. See E. Burns, *Bad Whisky*, Angel's Share, Glasgow, 2009.
10. A. MacDonald, *Whisky*, The Porpoise Press, Edinburgh, 1930.
11. S. H. Hastie, *From Burn to Bottle, Scotch Whisky Association*, Edinburgh, 1951.

CHAPTER 8

Today's Global Whisky Market

To put the data into some context, it may be helpful briefly to consider the performance of other spirits. The global consumption of
all spirits recorded as taxable is somewhere in excess of 2.5 *billion*
cases and, until the advent of the global economic recession, had
been growing at around 1.5% annually since the year 2000. To
that total, it is estimated that there are another 1.5 billion cases of
unrecorded or untaxed spirits, suggesting that annual global consumption of distilled spirits exceeds 4.0 billion cases.

This, of course, includes such products as aguardiente (Cachaça,
Guaro and Kleren are all Latin American or Caribbean variants on
aguardiente), Arrack, Baijiu, Horilka, Pisco, Shochu, Soju, fruit
brandies and schnapps, moonshine and other distilled spirits, all of
which may be of great significance in one particular area but do not
feature in world markets. It also includes all Indian spirits and,
while comprehensive or reliable data for India is hard to obtain, it
is suggested that sales of Indian spirits total in excess of 270 million
cases annually, of which perhaps half or more is Indian 'whisky'.

As a point of comparison, whatever the exact size of the domestic market for Indian 'whisky', it alone is certainly larger than
the *global* market for Scotch and US whiskies combined, which
goes some way to explaining the sustained efforts by the SWA and
DISCUS to eliminate Indian tariff barriers to imported whiskies.

The Science and Commerce of Whisky
By Ian Buxton and Paul S. Hughes
© Buxton & Hughes, 2014
Published by the Royal Society of Chemistry, www.rsc.org

The best-selling Indian brands, such as Officer's Choice, McDowells No. 1 and Bagpiper, each claim annual sales of 16 million cases or more.

Global vodka sales, including the Commenwealth of Independent States (CIS) countries, probably exceed 500 million cases; however, this falls to around 170 million cases when the CIS are excluded. Nonetheless, this means that total non-CIS vodka sales are higher than all whisky/whiskey (excluding Indian) sales. On an equivalent basis, global rum sales are around 135 million cases. Both rum and vodka (excluding the CIS) have recently exhibited growth rates in excess of total whisky.

The global whisky market has, in the past, been susceptible to periods of dramatic boom and bust, driven by such widely varying factors as general economic conditions; prohibition; political sanctions and grossly over-optimistic levels of production.

In addition, it should be stressed that the whisky market does not exist in isolation and is subject to competition from other spirits and pressures on discretionary leisure expenditure as a whole. Fashion plays a part in determining consumer choices and, as noted above, vodka in particular has been markedly successful over the last two decades in dramatically increasing its global appeal.

White spirits (gin, vodka and white rum primarily at an international level, although regionally other categories can be important) enjoy a significant competitive advantage in that they do not require any aging. Production of such products can therefore respond very quickly to changes in demand; whereas whisky suffers from an inevitable lag as a result of the requirement to age the product. Aging also involves a cost penalty as it requires casks, warehouse space and incurs evaporation losses, all of which place whisky at a competitive disadvantage. Management of these stocks is a complex logistical task, largely undertaken today by sophisticated computerised stock control systems (the romantically-minded reader may lament the passing of stencilled cask ends and their replacement by plastic strips carrying a bar code, but such is the price of tighter control and a digital system of information management).

As a measure of the physical scale of the task consider that Diageo's Blackgrange Bond in Alloa has some 3 300 000 casks in warehouses at any one time and aims to achieve space utilisation in excess of 99.4%, representing space available for just 6 casks in every 1000. This is just one of the company's sites and this is just one company, albeit the largest in the industry. Thus, it may be seen that the management of stock is a critically important, if largely unsung task and one which is closely linked to the equally complex task of sales forecasting.

The marketer of a 25 year old whisky in 2013 is dependent on decisions made in 1988 or earlier, under very different market conditions by executives highly unlikely to see the consequences of their decisions. And, to be able to market a 25 year old product in 2038, stocks will need to be laid down and reserved today on behalf of a future generation of management that those taking the initial decision are equally unlikely to know. At an extreme level, the distiller may very well have retired or even died before his product reaches the market – the producer of a 50 year old whisky will be faced with the organisational challenge of turning their inventory only twice in a century!

In practice, therefore, and with a few notable exceptions, the availability of really old whisky (over 40 years of age) has, until recently, been largely serendipitous and relied perhaps more heavily than the industry would like to admit on the 'lost cask' of marketing legend. The fairly widespread availability of aged Scotch whisky during the period 1990–2010 was an unforeseen if beneficial consequence of the 'Whisky Loch' and industry attitudes to aged whisky changed markedly during that period due, in part, to the availability of stock and, in part, to ready consumer acceptance of this whisky, marketed at premium prices. However, subsequent cutbacks in production are now reflected in the hasty development of high-priced non-age statement expressions reflecting continued demand for premium whiskies in a number of markets, many of which would not have featured in the sales planning of the 1970s and 1980s.

History teaches us that the consequences of poor planning can be catastrophic. While the Irish distilling industry dominated whisky sales prior to the late 19th century, changes in consumer taste and rigid management and political factors, combined with the dire effects of prohibition, all but wiped out Irish whiskey. It is only in the comparatively recent past that Irish whiskey has enjoyed the renaissance that is presently underway.

More recently, massive over-production in the late 1970s led to Scotland's 'Whisky Loch' of the mid-1980s, with attendant industry cutbacks and distillery closures. The loss of Port Ellen,

Rosebank and other distilleries is today lamented by whisky enthusiasts but such casualties were a consequence of irrational exuberance and over-confidence and an undue reliance on the single US market.

A major feature of today's global whisky market is that it is just that: *global*. While, hitherto, only Scotch has enjoyed significant international penetration; Irish and Canadian whiskies have had limited US distribution in addition to their domestic markets and Japanese and American whiskies have been largely confined to domestic sales; however, it is now the case that all of these whiskies and more from the new 'world' whisky nations can be found in many markets, at least in specialist outlets.

Having said that, according to the latest data supplied by just-drinks and the International Wine & Spirit Research,[1] non-Scotch whiskies still tend to be markedly less international in scope than other major international categories, such as Scotch or vodka. US whiskey sells more than a million cases in just four markets (the US, Germany, Australia and, most recently, the UK); Irish whiskey sells in just one market (the USA); Canadian whisky sells in two markets (Canada and the USA); Japanese whisky sells in one market (Japan); and Indian whisky sells in two markets (India and the UAE).

Vodka, by comparison, sells more than one million cases in more than 30 markets (including travel retail), Scotch in 22 markets and brandy in 21 markets.

So, there are many more markets for whisky, particularly Scotch whisky, than was historically the case. For some years following World War II, the USA accounted for more than 50% of Scotch whisky exports. An over-reliance on this single market proved highly problematic when, by the late 1970s, vodka and light rums grew rapidly in popularity there. As we shall see, the global market for whisky is now more widely spread, one factor in the increasing confidence of the industry in the reliability of its forecasts and demand modelling. What is seen as "*the democratisation of luxury purchasing*" is becoming a global norm and encouraging consumers to trade up to imported whiskies – it should be remembered that context is everything and thus a Londoner's bottle of standard blended Scotch (Johnnie Walker Red Label, for example) may well be seen as a prestigious luxury item to a resident of Mumbai or Bogota used to consuming locally produced spirits.

Considered in terms of Maslow's classic 'hierarchy of needs',[2] the move to premium spirits and the self-actualisation that they symbolise is part of the aspirational consumer's gradual progression from the satisfaction of basic needs to the fulfilment of higher ones, exemplified in the values of the brand. Such brands are seen as an affordable luxury that it is worth paying more for because they represent quality and value and offer status and peer group approval.

Based on these trends and the forecasts flowing from them, many commentators have concluded that whisky in general, and Scotch whisky in particular, is enjoying a 'golden age'. Certainly this has been reflected in very considerable investment in distilling capacity as mothballed distilleries (Glenglassaugh, Tamdhu, Glen Keith) have been re-opened; existing distilleries (Glenlivet, Muir of Ord and others) have been expanded and new distilleries (Roseisle, New Imperial) opened. In addition, Diageo have announced plans that may involve the construction of not one but potentially three new distilleries of similar scale to Roseisle (twelve million plus ola pa), *"if the annual sales growth of recent years is sustained for the next three or four years"*.

The development of a second Diageo distillery in two to three years' time (*i.e.*, 2014–2015) means that the company's total group malt whisky production will increase by 30–40%. Chivas Brothers, the Scotch whisky arm of Pernod Ricard, have also announced significant plans for expansion, on the site of the old Imperial distillery and elsewhere. In addition to the very substantial increase in capacity implied by these announcements, there are a number of smaller developments underway which, if nothing else, will add variety to the specialist sector and emphasise broad entrepreneurial confidence in the industry's future.

A similar phenomenon may be observed in the USA with an explosion of craft distillers. More significantly, at least in terms of volume, plans have been announced to construct a new $12m distillery in Louisville, Kentucky (the heart of the bourbon-producing region) for the Angel's Envy brand and Brown – Forman will spend $36m to double production at Woodford Reserve.

The Wild Turkey distillery was significantly expanded in 2011 and Maker's Mark aims to increase distilling and warehousing capacity by 50% by 2016. In total, the Kentucky Distillers Association estimate that around $150m has been spent in recent years in distillery expansion and upgraded visitor facilities, as bourbon positions itself to share in the global growth in whisk(e)y sales.

There are a number of factors behind this. These include renewed fashionability in developed markets, partly driven by growing interest in cocktails; a desire for imported 'status' brands in developing markets amongst their rapidly emerging and newly affluent middle class consumers; increased globalisation; more sophisticated marketing than was hitherto the case; consumer engagement with products that are seen to combine status with heritage, provenance and authenticity; whisky's new-found appeal to women, especially younger women, and increased product innovation. In addition, reduction of duties and the elimination of other non-tariff barriers to trade have further driven growth in markets such as South Korea, China and, to a lesser extent, India.

Conversely, and as a result of the growing confidence in the industry, there is an increasing view that whisky was for much of the past two decades considerably under-priced relative to competitor spirits. For many years the idea of whisky retailing at much over £100 per bottle was hard for the industry, retailers and consumers to countenance, let alone for this mark to be achieved.

However, once this barrier had been overcome, the upward movement of price became more rapid and whiskies were, by the early part of the present century, touching the £1000 mark for rare and super-premium expressions. More recently, that has been left far behind and, while hardly commonplace, single bottles retailing at £10 000 and more are thought unremarkable (Figure 8.1).

A few spectacular, show-piece bottles have achieved prices in excess of £100 000 per bottle: these include releases from The Dalmore and Bowmore. Blended whiskies have not been left behind this trend, with expressions from both Chivas Brothers' Royal Salute and Johnnie Walker both exceeding the £100 000 price point.

At the time of writing, the most expensive whiskies currently on offer are contained in a bespoke display cabinet in Harrods of London – the 12 bottle Richard Paterson Collection of The Dalmore, which is offered at £987 500 (Figure 8.2). The retailer has been quoted as saying that they are "*more concerned that we sell it too quickly and. . . lose an outstanding talking point and attraction in the spirits room*"! The collection attracted considerable media attention and much heated debate in blogging circles – which was, of course, a key part of the exercise for the brand.

Figure 8.1 Highland Park's 50 Year Old release – an example of super-premium whisky retailing at around £10 000.

Figure 8.2 Richard Paterson with the Dalmore Paterson Collection Cabinet on display at Harrods of London, which retails at £987 500 (presumably, delivery is included).

However, such whiskies, however remarkable and newsworthy they may be, constitute only a tiny part of the market. While marketers look to them to reflect glamour and prestige on the rest of the brand range, they are essentially PR exercises – albeit achieving real and highly profitable sales.

The majority of the whisky market exists in less rarefied conditions.

8.1 THE UK MARKET FOR WHISKY

Split between on-trade (all premises where alcohol is purchased and consumed 'on' the physical premises, such as pubs, hotels and restaurants) and off-trade (where alcohol is purchased for consumption 'off' the site, *i.e.*, supermarkets and wine and spirit merchants), the UK market for all whisky is estimated[3] to be worth £927m (up 2% by value) in the on-trade and nearly £1.2bn (+2% by value) in the off-trade.

Within the on-trade, sales of 1.09m cases are split as follows:

Blended Whisky	£425m	45.8% share by value
American Whiskey	£340m	36.6% share by value
Malt Whisky	£106m	11.4% share by value
Irish Whiskey	£50m	5.4% share by value
Other (*e.g.*, Canadian)	£7m	0.8% share by value

The importance of American whiskey in the on-trade contrasts strongly with its position in off-trade sales (see the text that follows).

While there are some eight million blended Scotch drinkers in the UK on-trade, a third are over 65 years old, contrasting with malt whisky (more than half – 54% – under the age of 55); Irish (one million drinkers; more than 40% under the age of 45) and American (3.6m drinkers; 43% under the age of 35 – by some distance, the youngest demographic).

The UK's top five blended whiskies in the on-trade are:

1. The Famous Grouse
2. Bell's Original
3. Johnnie Walker Black Label 12 Year Old
4. Whyte & Mackay
5. Chivas Regal 12 Year Old

The UK's top five American whiskies in the on-trade are:

1. Jack Daniel's
2. Maker's Mark
3. Jim Beam
4. Bulleit Bourbon
5. Jack Daniel's Gentleman Jack

The UK's top five malt whiskies in the on-trade are:

1. Glenfiddich 12 Year Old
2. Laphroaig 10 Year Old
3. Glenmorangie 10 Year Old
4. Macallan Sherry Oak 10 Year Old
5. Highland Park 12 Year Old

Within the off-trade, sales of 6.3m cases are split as follows:

Blended Whisky	£824m	71% share by value
Malt Whisky	£160m	14% share by value
American Whiskey	£144m	12% share by value
Irish Whiskey	£35m	3% share by value

75% of all UK on-trade whisky drinkers are male, with Irish and American whiskeys having a dramatically younger demographic profile. Purchases for gifting are particularly significant for the sales of malt and Irish whiskies.

The UK's top five blended whiskies in the off-trade are:

1. The Famous Grouse
2. Bell's Original
3. Grant's Family Reserve
4. Whyte & Mackay
5. High Commissioner

The UK top five malt whiskies in the off-trade are:

1. Glenfiddich 12 Year Old
2. Glenmorangie Original

3. Isle of Jura 10 Year Old
4. The Glenlivet 12 Year Old
5. Aberlour 10 Year Old

Flavoured American whiskies, such as Jack Daniel's Tennessee Honey and others, are anticipated to represent a major growth sector in the near future.

8.2 SCOTCH WHISKY – GLOBAL

The market is divided into blended Scotch and malt Scotch. Blends remain easily the largest component. Global blends reached 88.0m cases in 2012 (including duty free and domestic), up from 80.6m cases in 2006. The fast recovery in volumes after the 2008/2009 global financial crisis is especially noteworthy (Table 8.1).

Europe remains the largest region, despite being the only major region to post a decline over the period (Table 8.1). France is easily the largest market for blended Scotch, with sales of 13.18m cases, much sold at relatively low prices. The USA is the next-largest market at 6.9m cases, although this is projected to decline slowly (Table 8.2).

Despite attracting much commentary, especially in the UK and USA, single malt sales still only account for some 8% of total Scotch sales, at 7.8m cases (Tables 8.3 and 8.4). Growth over the period has been strong however (+ 28.3% *vs.* 2006) and is forecast to continue. That said, it should be noted that many brands are constrained either by production capacity or by the demands of

Table 8.1   Global blended Scotch whisky – volume by major region.

	2006	*2007*	*2008*	*2009*	*2010*	*2011*	*2012*
Total	80 656.5	84 123.9	83 428.1	80 606.2	83 043.7	85 909.4	88 010.0
Europe	36 782.5	37 069.3	36 502.9	35 401.4	34 899.0	34 262.6	33 813.5
Americas	19 986.8	21 115.8	20 686.4	19 898.6	20 575.8	21 885.1	22 515.0
Asia Pacific	13 049.4	13 260.1	12 843.7	12 344.0	12 707.4	13 378.8	13 772.3
Africa & Middle East	5378.7	6625.0	6756.4	6985.3	7441.3	7541.9	7722.3
Travel Retail	4321.5	4444.7	4604.7	4075.5	4823.6	5204.0	5470.0
CIS	779.3	1246.2	1608.3	1510.8	2205.8	3246.3	4326.2
Rest of World	358.4	362.9	425.9	390.9	390.9	390.8	390.8

Source: The IWSR; just-drinks.

Table 8.2   Blended Scotch whisky – top 25 markets by volume.

	2006	2007	2008	2009	2010	2011	2012
Total	80 656.5	84 123.9	83 428.1	80 606.2	83 043.7	85 909.4	88 010.0
France	11 093.3	11 819.0	12 209.0	12 656.8	12 969.8	13 185.3	13 216.8
United States	7526.0	7515.0	7410.5	7034.3	6870.0	6935.0	6875.0
United Kingdom	6254.8	6061.0	5899.9	5852.1	5805.2	5478.3	5373.5
Travel Retail	4321.5	4444.7	4604.7	4075.5	4823.6	5204.0	5470.0
Spain	7476.3	7136.5	6415.3	5527.8	5058.2	4539.5	4189.0
Brazil	2565.5	2797.5	2844.5	2840.8	3468.0	4085.8	4340.9
Russia	633.8	1049.6	1399.1	1306.5	1939.5	2864.9	3861.3
Thailand	3865.5	3673.0	3219.3	2984.5	2810.9	2747.6	2627.7
Mexico	1229.0	1451.8	1694.5	1848.3	2233.3	2591.5	2910.3
South Africa	1907.3	2238.0	2167.5	2062.8	2323.0	2480.3	2563.3
South Korea	2656.5	2751.3	2756.0	2451.0	2413.0	2289.8	2203.8
India	652.5	963.5	1165.3	1384.5	1555.0	1958.5	2310.0
Ven ezuela	2879.3	3039.1	2512.0	2396.1	1982.9	1872.8	1800.0
China	1557.8	1760.0	1755.8	1572.3	1637.3	1681.8	1687.8
Australia	1816.8	1651.5	1586.5	1549.5	1542.1	1610.8	1674.3
Colombia	1191.5	1474.5	1330.1	1267.0	1347.5	1497.2	1618.8
Germany	1548.8	1489.5	1484.5	1418.5	1435.0	1444.8	1459.8
Greece	2456.5	2400.0	2332.5	2289.3	1741.0	1372.3	1127.8
Poland	325.8	446.5	556.0	606.7	824.8	1126.8	1357.8
Portugal	1422.3	1333.8	1263.3	1132.0	1104.3	1048.3	969.3
Japan	903.3	826.5	723.5	692.3	729.0	817.3	827.5
Belgium and Luxembourg	772.2	794.9	760.9	759.5	761.8	784.0	789.0
Netherlands	684.9	714.3	662.8	641.5	689.8	715.5	727.0
Taiwan	611.3	577.3	540.8	529.3	641.3	689.8	696.3
Italy	920.8	812.0	714.0	636.5	682.3	670.0	645.5
Others	13 383.9	14 903.5	15 420.2	15 091.4	15 655.6	16 218.3	16 688.2

Source: The IWSR; just-drinks.

blending. It is not unknown for single malt brands to be 'on allocation' where demand exceeds the available supply.

Examining Table 8.5, several points are immediately apparent. Firstly, the dominant position enjoyed by Diageo's Johnnie Walker, which accounts for more than 17.5% of all Scotch whisky sales, is more than twice as large as the entire single malt category and more than two and half times larger than its nearest competitor. Diageo also have six other brands each selling more than one million cases. Secondly, the strong influence of French-owned groups, with some 21.8m cases of sales, and finally the appearance of Glenfiddich, the only single malt to enjoy annual sales of more than one million cases.

Table 8.3 Global malt Scotch whisky – volume by major region.

	2006	2007	2008	2009	2010	2011	2012
Total	6109.6	6423.5	6423.9	6133.9	6707.1	7356.2	7840.6
Europe	3117.6	3199.0	3054.6	2798.2	2904.5	2965.9	2994.8
Americas	1183.7	1264.7	1309.4	1311.1	1412.8	1584.1	1717.0
Asia Pacific	903.3	961.6	1011.7	1010.4	1182.3	1420.2	1583.3
Travel Retail	745.4	801.0	812.2	800.6	945.1	1078.2	1194.0
Africa & Middle East	108.6	119.8	132.6	130.9	150.5	169.0	186.1
CIS	31.3	53.7	70.1	45.5	74.6	101.1	127.8
Rest of World	19.9	24.0	33.3	37.4	37.4	37.7	37.7

Source: The IWSR; just-drinks.

Table 8.4 Malt Scotch whisky – top 25 markets by volume.

	2006	2007	2008	2009	2010	2011	2012
Total	6109.6	6423.5	6423.9	6133.9	6707.1	7356.2	7840.6
United States	954.0	1015.0	1050.3	1055.8	1135.3	1260.0	1360.0
Travel Retail	745.4	801.0	812.2	800.6	945.1	1078.2	1194.0
France	898.8	911.3	857.5	811.5	805.5	840.3	844.8
United Kingdom	713.3	740.8	738.6	723.9	806.9	826.8	843.5
Taiwan	599.0	599.8	603.0	547.0	660.5	773.3	861.5
Germany	277.0	278.0	268.0	270.0	295.0	320.3	340.3
Italy	428.8	442.8	385.0	274.5	274.3	251.0	233.5
Canada	180.5	190.3	195.8	194.8	207.8	232.0	255.0
Japan	146.5	165.3	154.8	150.8	162.5	171.5	173.0
Spain	308.3	291.8	246.5	189.0	177.0	159.0	144.3
Sweden	105.0	115.0	129.7	126.9	132.4	130.0	129.8
India	18.3	23.0	33.5	54.5	70.0	112.8	145.0
Netherlands	80.2	88.3	83.3	78.0	86.5	101.5	111.3
Australia	51.8	58.5	65.0	65.5	71.7	89.3	93.0
China	33.0	40.3	59.3	69.3	66.5	88.8	99.5
Russia	26.0	47.1	61.2	36.3	60.5	80.0	102.3
South Africa	42.5	54.5	55.3	55.0	60.0	67.8	75.3
Belgium and Luxembourg	55.3	54.1	55.9	53.5	58.8	63.3	66.8
Switzerland	49.0	51.5	57.9	59.3	59.8	59.3	60.8
South Korea	22.8	34.0	40.5	47.3	51.5	57.8	61.5
Malaysia	3.5	7.0	12.2	19.3	30.6	41.9	49.5
Brazil	12.5	15.8	17.5	12.5	23.3	39.3	45.1
Greece	57.8	65.8	67.3	58.8	35.8	29.0	21.3
Vietnam	1.9	2.8	6.5	15.2	18.7	28.8	36.1
Austria	21.0	21.1	21.8	23.5	26.9	28.3	29.4
Others	277.9	309.4	345.1	341.7	384.7	426.6	464.5

Source: The IWSR; just-drinks.

Table 8.5 Scotch whisky brands with sales in excess of one million cases – 2011.

Brand	Owner	Est. Sales – million cases[a].*
Johnnie Walker	Diageo	16.9
Ballantine's	Pernod Ricard	6.47
Grant's Founder's Reserve	William Grant & Sons	4.97
Chivas Regal	Pernod Ricard	4.89
J & B Rare	Diageo	4.80
Dewar's	Bacardi	3.19
The Famous Grouse	Edrington	3.00
William Peel	Belvédère	2.90
Bell's	Diageo	2.50
Label 5	La Martiniquaise	2.50
William Lawson's	Bacardi	2.29
Teacher's	Beam	2.05
Clan Campbell	Pernod Ricard	1.96
100 Pipers	Pernod Ricard	1.71
Buchanan's	Diageo	1.60
Sir Edward's	La Martiniquaise	1.35
White Horse	Diageo	1.25
Clan MacGregor	William Grant & Sons	1.14
Black & White	Diageo	1.05
Old Parr	Diageo	1.05
Glenfiddich	William Grant & Sons	1.03

[a]Source: Drinks International Millionaires Club/IWSR/just-drinks/authors' estimates.
*Now reported to exceed 20 million cases.

8.3 OTHER WHISKY

This category includes American, Canadian, Irish, Indian and Japanese whiskies and, as the Table 8.6 makes clear, is dominated by Indian 'whisky' – the majority of which, as the reader will be aware, does not qualify as whisky by international definitions. However, although a small amount of whisky meeting the definitions is now made in India the volumes are, by global standards, insignificant.

It is worth noting the remarkable growth in 'other whisky', the majority of which has been driven by domestic Indian demand for Indian whiskies, which are also popular in the UAE.

The revival in Irish whiskey, albeit from a relatively small base, is also marked. With increased investment by the major brand, Jameson, and ambitious new entrants, such as William Grant & Sons with their new Tullamore Dew distillery and Beam's purchase

Table 8.6 Other whiskies – global volumes in '000s of cases.

	2007	2008	2009	2010	F2011
Total (non-Scotch)	148 471.8	171 069.9	189 941.7	209 614.5	219 458.1
US Whiskey	28 982.27	29 522.40	29 423.32	29 983.24	29 233.73
Canadian Whisky	20 692.33	20 879.72	20 439.79	20 273.25	19 863.50
Irish Whiskey	4097.93	4452.12	4439.73	4952.57	4934.60
Other Whisky	94 699.32	116 215.66	135 638.86	154 405.40	165 426.28

Source: The IWSR; just-drinks.

of Cooley, the category appears to be set for further dynamic growth. Major markets for Irish whiskey are the USA, where Jameson is dominant. In December 2010, Pernod Ricard announced that Jameson, the industry flagship, had surpassed three million nine-litre cases for the first time.

Pernod's Irish operating company, Irish Distillers Ltd (IDL), has also innovated with some success in bringing back the traditional pot still style of Irish whiskey, which had previously fallen out of favour. Brands in this category appear capable of taking sales from single malt Scotch.

As with Scotch and bourbon, there has been a considerable investment in expanded capacity – Bushmills has been doubled to an output of one million cases annually; William Grant & Sons are building a brand new distillery for Tullamore Dew at a reported cost of £35m; having sold Cooley, the Teeling family have re-invested the proceeds in a new operation and expansion at Jameson's Midleton distillery, near Cork, has been very substantial indeed.

In addition, IDL have invested around €100m in a new distillery on a green-field site at Dungourney, which will come on-stream late in 2013.

8.4   IN CONCLUSION...

In conclusion, and at the risk of tempting fate, the overall state and future prospects for the global whisky market and its various sub-sectors seem as positive as they have been at any time in the past.

With improvements in production technology leading to ever more sustained product quality and consistency, enhanced environmental practices, improved marketing and sales strategies, greater consumer knowledge, affluence and demand for premium products of authenticity and provenance the future looks bright.

The next generation of management of the whisky distilling industry, while holding hundreds of years of history and irreplaceable heritage in their hands, have a once-in-a-lifetime opportunity to build and develop the industry, taking it to new levels of global success.

As Robert Burns once wrote, albeit in a different context: *"freedom and whisky gang thegither"*. It is our hope that our successors in whisky succeed in taking their freedom to develop whisky in all its forms to unparalled heights. Let this be their motto:

"Too much of anything is bad, but too much of good whiskey is barely enough".

Mark Twain

REFERENCES

1. A. Smith, The IWSR/just-drinks reports, "Global Market Review of Blended & Single Malt Scotch Whisky - Forecasts to 2017" and "Global Market Review of non-Scotch Whiskies", Aroq Ltd and The IWSR, Bromsgrove, 2013/2011.
2. A. Maslow, *Motivation and Personality*, Harper, New York, 1954.
3. First Drinks Brands, *First Drinks Market Report 2013*, Hook, 2013.

Subject Index